Delmar Learning's Test Preparation Series

School Bus Test

Brakes (Test S4)

THOMSON

DELMAR LEARNING

Australia Canada Mexico Singapore Spain United Kingdom United States

THOMSON
™
DELMAR LEARNING

Delmar Learning's ASE Test Preparation Series

School Bus Test for Brakes (S4)

Vice President, Technology and Trades SBU:
Alar Elken

Executive Director, Professional Business Unit:
Greg Clayton

Product Development Manager:
Timothy Waters

Development:
Kristen Shenfield

Channel Manager:
Beth A. Lutz

Marketing Specialist:
Brian McGrath

Production Director:
Mary Ellen Black

Production Manager:
Larry Main

Production Editor:
Tom Stover

Editorial Assistant:
Kristen Shenfield

Cover Designer:
Michael Egan

NOTICE TO THE READER

Contents

Section 1 The History of ASE

Section 2 Take and Pass Every ASE Test

Section 3 Types of Questions on an ASE Exam

Section 4 Overview of the Task List

Section 5 Sample Test for Practice

Section 6 Additional Test Questions for Practice

Section 7 Appendices

Preface

Delmar Learning is very pleased that you have chosen our ASE Test Preparation Series to prepare yourself for the school bus ASE Examinations. These guides are available for the school bus areas S2, S4, and S5. These guides are designed to introduce you to the Task List for the test you are preparing to take, give you an understanding of what you are expected to be able to do in each task, and take you through sample test questions formatted in the same way the ASE tests are structured. If you have a working knowledge of the discipline you are testing for, you will find the Delmar ASE Test Preparation Series to be an excellent way to understand the "must know" items to pass the test. These books are not textbooks. Their objective is to prepare the technician who has the requisite experience and schooling to challenge ASE testing. It cannot replace the hands-on experience or the theoretical knowledge required by ASE to master vehicle repair technology. If you are unable to understand more than a few of the questions and their explanations in this book, it could be that you require either more shop-floor experience or further study. Some textbooks to assist you with further study are listed on the rear cover of this book.

Each book begins with an item-by-item overview of the ASE Task list with explanations of the minimum knowledge you must possess to answer questions related to the task. Following that there are two sets of sample questions followed by an answer key to each test and an explanation of the answers to each question. A few of the questions are not strictly ASE format but were included because they help to teach a critical concept that will appear on the test. We suggest that you read the complete Task List Overview before taking the first sample test. After taking the first test, score yourself and read the explanation to any questions that you were not sure about, including the questions you answered correctly. Each test question has a reference back to the related task or tasks that it covers. This will help you to go back and re-read any area of the task list that you are having trouble with. Once you are satisfied that you have all of your questions answered from the first sample test take the second one and check it. If you pass these tests, you will do well on the ASE test.

Our Commitment to Excellence

This edition of Delmar Learning's ASE Test Preparation Series has been updated to ASE's task lists and test questions and has been checked for accuracy. Delmar Learning has sought out the best technicians in the country to help with the updating and revision of each of the books in the series.

About the Author

Sean Bennett has an industry background with a major truck corporation in Allentown, PA, and currently coordinates truck technology training for a truck company in Toronto. He has held ASE Master (Truck) Certification and taught at both college and corporate training levels. Additionally, he is the author of a number of Delmar publications including *Truck Engine, Fuel and Computerized Management Systems, Heavy Duty Truck Systems, Truck Diesel Engines,* and in addition to acting as the series advisor for *ASE Test Preparation for Medium Heavy-Duty Trucks T1–T8,* 3e, he was also the revision author for book T6 in the series. He has made a lifelong commitment to adult education and believes that programs emphasizing performance-based learning outcomes are critical for student success in transportation technology education.

Thanks for choosing Delmar Learning's ASE Test Preparation Series. All of the writers, editors, Delmar Learning staff, and myself have worked very hard to make this series second to none. I know you are going to find this book accurate and easy to work with. It is our objective to constantly improve our product at Delmar by responding to feedback. If you have any questions concerning the books in this series you can email me at truckexpert@trainingbay.com.

Sean Bennett
Series Advisor

The History of ASE

History

Originally known as The National Institute for Automotive Service Excellence (NIASE), today's ASE was founded in 1972 as a nonprofit, independent entity dedicated to improving the quality of automotive service and repair through the voluntary testing and certification of automotive technicians. Until that time, consumers had no way of distinguishing between competent and incompetent automotive mechanics. In the mid-1960s and early 1970s, efforts were made by several automotive industry affiliated associations to respond to this need. Though the associations were nonprofit, many regarded certification test fees merely as a means of raising additional operating capital. Also, some associations, having a vested interest, produced test scores heavily weighted in the favor of its members.

From these efforts a new independent, nonprofit association, the National Institute for Automotive Service Excellence (NIASE), was established. In early NIASE tests, Mechanic A, Mechanic B type questions were used. Over the years the trend has not changed, but in mid-1984 the term was changed to Technician A, Technician B to better emphasize sophistication of the skills needed to perform successfully in the modern motor vehicle industry. In certain tests the term used is Estimator A/B, Painter A/B, or Parts Specialist A/B. At about that same time, the logo was changed from "The Gear" to "The Blue Seal," and the organization adopted the acronym ASE for Automotive Service Excellence.

ASE

ASE's mission is to improve the quality of vehicle repair and service in the United States through the testing and certification of automotive repair technicians. Prospective candidates register for and take one or more of ASE's many exams.

Upon passing at least one exam and providing proof of two years of related work experience, the technician becomes ASE certified. A technician who passes a series of exams earns ASE Master Technician status. An automobile technician, for example, must pass eight exams for this recognition.

The exams, conducted twice a year at over seven hundred locations around the country, are administered by American College Testing (ACT). They stress real-world diagnostic and repair problems. Though a good knowledge of theory is helpful to the technician in answering many of the questions, there are no questions specifically on theory. Certification is valid for five years. To retain certification, the technician must be retested to renew his or her certificate.

The automotive consumer benefits because ASE certification is a valuable yardstick by which to measure the knowledge and skills of individual technicians, as well as their commitment to their chosen profession. It is also a tribute to the repair facility employing ASE certified technicians. ASE certified technicians are permitted to wear blue and white ASE shoulder insignia, referred to as the "Blue Seal of Excellence," and carry credentials listing their areas of expertise. Often employers display their technicians' credentials in the customer waiting area. Customers look for facilities that display ASE's Blue Seal of Excellence logo on outdoor signs, in the customer waiting area, in the telephone book (Yellow Pages), and in newspaper advertisements.

To become ASE certified, contact:

National Institute for Automotive Service Excellence
101 Blue Seal Drive S.E.
Suite 101
Leesburg, VA 20175
Telephone 703-669-6600
FAX 703-669-6123
www.ase.com

2 Take and Pass Every ASE Test

ASE Testing

Participating in an Automotive Service Excellence (ASE) voluntary certification program gives you a chance to show your customers that you have the "know-how" needed to work on today's modern vehicles. The ASE certification tests allow you to compare your skills and knowledge to the automotive service industry's standards for each specialty area.

If you are the "average" automotive technician taking this test, you are in your mid-thirties and have not attended school for about fifteen years. That means you probably have not taken a test in many years. Some of you, on the other hand, have attended college or taken postsecondary education courses and may be more familiar with taking tests and with test-taking strategies. There is, however, a difference in the ASE test you are preparing to take and the educational tests you may be accustomed to.

Who Writes the Questions?

The questions, written by service industry experts familiar with all aspects of service consulting, are entirely job related. They are designed to test the skills that you need to know to work as a successful technician; theoretical knowledge is not covered.

Each question has its roots in an ASE "item-writing" workshop where service representatives from automobile manufacturers (domestic and import), aftermarket parts and equipment manufacturers, working technicians, and vocational educators meet in a workshop setting to share ideas and translate them into test questions. Each test question written by these experts must survive review by all members of the group. The questions are written to deal with practical application of soft skills and product knowledge experienced by technicians in their day-to-day work.

All questions are pretested and quality-checked on a national sample of technicians. Those questions that meet ASE standards of quality and accuracy are included in the scored sections of the tests; the "rejects" are sent back to the drawing board or discarded altogether.

Each certification test is made up of between forty and eighty multiple-choice questions. The testing sessions are 4 hours and 15 minutes, allowing plenty of time to complete several tests.

Note: Each test could contain additional questions that are included for statistical research purposes only. Your answers to these questions will not affect your score, but since you do not know which ones they are, you should answer all questions in the test. The five-year Recertification Test will cover the same content areas as those listed above. However, the number of questions in each content area of the Recertification Test will be reduced by about one-half.

Objective Tests

A test is called an objective test if the same standards and conditions apply to everyone taking the test and there is only one correct answer to each question. Objective tests primarily measure your ability to recall information. A well-designed objective test can also test your ability to understand, analyze, interpret, and apply your knowledge. Objective tests include true-false, multiple choice, fill in the blank, and matching questions. ASE's tests consist exclusively of four-part multiple-choice objective questions.

Before beginning to take an objective test, quickly look over the test to determine the number of questions, but do not try to read through all of the questions. In an ASE test, there are usually between forty and eighty questions, depending on the subject. Read through each question before marking your answer. Answer the questions in the order they appear on the test. Leave the questions blank that you are not sure of and move on to the next question. You can return to those unanswered questions after you have finished the others. They may be easier to answer at a later time after your mind has had additional time to consider them on a subconscious level. In addition, you might find information in other questions that will help you to answer some of them.

Do not be obsessed by the apparent pattern of responses. For example, do not be influenced by a pattern like **D, C, B, A, D, C, B, A** on an ASE test.

There is also a lot of folk wisdom about taking objective tests. For example, there are those who would advise you to avoid response options that use certain words such as *all, none, always, never, must,* and *only,* to name a few. This, they claim, is because nothing in life is exclusive. They would advise you to choose response options that use words that allow for some exception, such as *sometimes, frequently, rarely, often, usually, seldom,* and *normally.* They would also advise you to avoid the first and last option (A and D) because test writers, they feel, are more comfortable if they put the correct answer in the middle (B and C) of the choices. Another recommendation often offered is to select the option that is either shorter or longer than the other three choices because it is more likely to be correct. Some would advise you to never change an answer since your first intuition is usually correct.

Although there may be a grain of truth in this folk wisdom, ASE test writers try to avoid them and so should you. There are just as many **A** answers as there are **B** answers, just as many **D** answers as **C** answers. As a matter of fact, ASE tries to balance the answers at about 25 percent per choice **A, B, C,** and **D.** There is no intention to use "tricky" words, such as outlined above. Put no credence in the opposing words "sometimes" and "never," for example.

Multiple-choice tests are sometimes challenging because there are often several choices that may seem possible, and it may be difficult to decide on the correct choice. The best strategy, in this case, is to first determine the correct answer before looking at the options. If you see the answer you decided on, you should still examine the options to make sure that none seem more correct than yours. If you do not know or are not sure of the answer, read each option very carefully and try to eliminate those options that you know to be wrong. That way, you can often arrive at the correct choice through a process of elimination.

If you have gone through all of the test and you still do not know the answer to some of the questions, then guess. Yes, guess. You then have at least a 25 percent chance of being correct. If you leave the question blank, you have no chance. In ASE tests, there is no penalty for being wrong.

Preparing for the Exam

The main reason we have included so many sample and practice questions in this guide is, simply, to help you learn what you know and what you don't know. We recommend that you work your way through each question in this book. Before doing this, carefully look through Section 3; it contains a description and explanation of the questions you'll find in an ASE exam.

Once you know what the questions will look like, move to the sample test. After you have answered one of the sample questions (Section 5), read the explanation (Section 7) to the answer for that question. If you don't feel you understand the reasoning for the correct answer, go back and read the overview (Section 4) for the task that is related to that question. If you still don't feel you have a solid understanding of the material, identify a good source of information on the topic, such as a textbook, and do some more studying.

After you have completed the sample test, move to the additional questions (Section 6). This time answer the questions as if you were taking an actual test. Once you have answered all of the questions, grade your results using the answer key in Section 7. For every question that you gave a wrong answer to, study the explanations to the answers and/or the overview of the related task areas.

Here are some basic guidelines to follow while preparing for the exam:

- Focus your studies on those areas you are weak in.
- Be honest with yourself while determining if you understand something.
- Study often but in short periods of time.
- Remove yourself from all distractions while studying.
- Keep in mind the goal of studying is not just to pass the exam, the real goal is to learn!

During the Test

Mark your bubble sheet clearly and accurately. One of the biggest problems an adult faces in test taking, it seems, is in placing an answer in the correct spot on a bubble sheet. Make certain that you mark your answer for, say, question 21, in the space on the bubble sheet designated for the answer for question 21. A correct response in the wrong bubble will probably be wrong. Remember, the answer sheet is machine scored and can only "read" what you have bubbled in. Also, do not bubble in two answers for the same question.

If you finish answering all of the questions on a test ahead of time, go back and review the answers of those questions that you were not sure of. You can often catch careless errors by using the remaining time to review your answers.

At practically every test, some technicians will invariably finish ahead of time and turn their papers in long before the final call. Do not let them distract or intimidate you. Either they knew too little and could not finish the test, or they were very self-confident and thought they knew it all. Perhaps they were trying to impress the proctor or other technicians about how much they know. Often you may hear them later talking about the information they knew all the while but forgot to respond on their answer sheet.

It is not wise to use less than the total amount of time that you are allotted for a test. If there are any doubts, take the time for review. Any product can usually be made better with some additional effort. A test is no exception. It is not necessary to turn in your test paper until you are told to do so.

Your Test Results!

You can gain a better perspective about tests if you know and understand how they are scored. ASE's tests are scored by American College Testing (ACT), a nonpartial, unbiased organization having no vested interest in ASE or in the automotive industry. Each question carries the same weight as any other question. For example, if there are fifty questions, each is worth 2 percent of the total score. The passing grade is 70 percent. That means you must correctly answer thirty-five of the fifty questions to pass the test.

The test results can tell you:

- where your knowledge equals or exceeds that needed for competent performance, or
- where you might need more preparation.

The test results *cannot* tell you:

- how you compare with other technicians, or
- how many questions you answered correctly.

Your ASE test score report will show the number of correct answers you got in each of the content areas. These numbers provide information about your performance in each area of the test. However, because there may be a different number of questions in each area of the test, a high percentage of correct answers in an area with few questions may not offset a low percentage in an area with many questions.

It may be noted that one does not "fail" an ASE test. The technician who does not pass is simply told "More Preparation Needed." Though large differences in percentages may indicate problem areas, it is important to consider how many questions were asked in each area. Since each test evaluates all phases of the work involved in a service specialty, you should be prepared in each area. A low score in one area could keep you from passing an entire test.

There is no such thing as average. You cannot determine your overall test score by adding the percentages given for each task area and dividing by the number of areas. It doesn't work that way because there generally are not the same number of questions in each task area. A task area with twenty questions, for example, counts more toward your total score than a task area with ten questions.

Your test report should give you a good picture of your results and a better understanding of your task areas of strength and weakness.

If you fail to pass the test, you may take it again at any time it is scheduled to be administered. You are the only one who will receive your test score. Test scores will not be given over the telephone by ASE nor will they be released to anyone without your written permission.

3 Are You Sure You're Ready for Test S4?

Pretest

The purpose of this pretest is to determine the amount of review that you may require prior to taking the Automative Service Excellence (ASE) school bus test: Brakes (Test S4). If you answer all of the pretest questions correctly, complete the sample test in Section 5 along with the additional test questions in Section 6.

If two or more of your answers to the pretest questions are wrong, study Section 4, An Overview of the System, before continuing with the sample test and additional test questions.

The pretest answers and explanations are located at the end of the pretest.

1. The compressor for a school bus air-braking system is driven by:
    A. a separate electrical motor.
    B. the vehicle's engine.
    C. the airflow moving past its vanes.
    D. heated air from the engine.

2. What is another name for the supply tank?
    A. Reservoir tank
    B. Wet tank
    C. Dry tank
    D. Primary reservoir

3. A bus equipped with air brakes has parking brake release problems. Technician A says the cause could be a defective spring brake chamber. Technician B says the parking brake cable may be seized. Who is right?
    A. A only
    B. B only
    C. Both A and B
    D. Neither A nor B

4. Technician A says that school buses equipped with hydraulic brakes must use a dual-circuit hydraulic system. Technician B says that school buses equipped with air brakes are required to be FMVSS 121 compliant. Who is right?
    A. A only
    B. B only
    C. Both A and B
    D. Neither A nor B

5. All of the following are part of a school bus air-brake system **EXCEPT:**
    A. drums.
    B. S-camshafts.
    C. reservoirs.
    D. rims.

6. All of the following are part of a school bus hydraulic-braking system **EXCEPT:**
   A. the master cylinder.
   B. the governor.
   C. the wheel cylinders.
   D. the metering valve.

7. Which would be the LEAST-Likely cause of an antilock brake system (ABS) failure?
   A. A failed wheel-speed sensor
   B. A failed electronic control unit
   C. A defective modulator assembly
   D. A chipped wheel-sensor tooth

8. Which of the following would be the LEAST-Likely cause of wheel lockup?
   A. Improperly adjusted brakes
   B. Leaking service chamber
   C. Bald tires
   D. Aggressive braking on slippery roads

9. The Most-Likely cause of low system air pressure in a bus air-brake system would be:
   A. out-of-adjustment governor.
   B. leaking primary reservoir.
   C. leaking treadle valve.
   D. defective ratio valve.

10. The Most-Likely result of a complete secondary circuit failure in a school bus air-brake system would be:
    A. full-wheel lockup.
    B. complete-brake failure.
    C. visible and audible driver alert.
    D. loss of parking brakes.

11. Technician A says that when an air-brake equipped school bus is running down the highway and the driver applies the service brakes, air pressure will be present in both the service and hold-off chambers in any spring brake assembly.
    Technician B says that that hold-off chamber air is exhausted when parking brakes are applied. Who is correct?
    A. A only
    B. B only
    C. Both A and B
    D. Neither A nor B

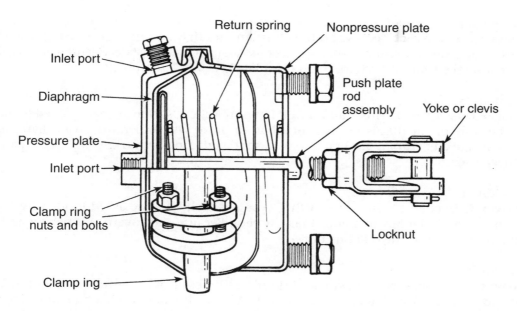

12. In the figure, observe the air-brake chamber. Technician A says that this assembly
    should never be disassembled without first caging the spring. Technician B says
    that this type of brake chamber is used only for parking-brake applications.
    Who is right?
    A. A only
    B. B only
    C. Both A and B
    D. Neither A nor B

Answers to the Test Questions for the Pretest

1. B, 2. B, 3. A, 4. C, 5. D, 6. B, 7. D, 8. B, 9. A, 10. C, 11. C, 12. D

Explanations to the Answers for the Pretest

Question #1
Answer B is correct. In this type of question, you are completing a sentence. The compressor for a school bus air-braking system is driven by the vehicle engine.

Question #2
Answer B is correct. You are simply answering a direct question; another name for the supply tank is a wet tank.

Question #3
Answer A is correct. In a Technician A, Technician B question, you are being asked to determine whether A is right, B is right, both are right, or both are wrong. Technician A says that a defective spring-brake chamber could cause parking brake release problems, and he is right because the hold-off chamber could be leaking. Technician B says that the problem could be due to a parking-brake cable, and he is wrong because bus air-brake systems have no parking-brake cable. Because Technician A is right and Technician B is wrong, answer A is the correct choice.

Question #4

Answer C is correct because both technicians are right. Technician A is right; school buses equipped with hydraulic-brakes must use a dual-circuit system. Technician B is also right; air-brake equipped school buses must be FMVSS 121 compliant.

Question #5

Answer D is correct. This is an **EXCEPT**-type question, so you are being asked to identify the component that is not part of a bus air-brake system. Drums, S-camshafts, and reservoirs are all part of a bus brake system, so the odd one out is rims, which are part of a wheel assembly.

Question #6

Answer B is correct. In this **EXCEPT**-type question, you are asked to identify components in a school bus hydraulic-brake system. Because the governor is part of an air-brake system, this is the odd one out.

Question #7

Answer D is correct. A failed wheel-speed sensor, electronic control unit, and modulator can all cause ABS system failure, so the LEAST-Likely cause would be a chipped tooth in a wheel-speed sensor.

Question #8

Answer B is correct. The LEAST-Likely cause of a wheel lockup condition would be a leaking service chamber, because service-braking application to the wheel would be lowered as a result.

Question #9

Answer A is correct. System air pressure is managed by the governor, so the Most-Likely cause would be an out-of-adjustment governor.

Question #10

Answer C is correct. The Most-Likely result of a complete secondary circuit failure in a school bus air-brake system would be that the driver would be alerted to the condition by audible (buzzer) and visible (warning light) means. School bus driver training dictates that the vehicle immediately pull over and await assistance.

Question #11

Answer C is correct because both technicians are right. Technician A understands how a spring-brake chamber functions, hold-off air is required to release the brake and so the hold-off chamber is charged with air whenever the vehicle park brakes are released. That means that during service braking, air will be charged to both the service and hold-off chambers. Technician B is also right. He understands that hold-off air is required to release a parking brake, that is, pneumatically cage the parking brake spring.

Question #12

Answer D is correct because both technicians are wrong. The figure shows a typical service chamber. Technician A is wrong; this unit has no power spring. The spring shown is a retraction spring and these units are often disassembled to replace the service diaphragm. Technician B is also wrong; this type of chamber is air actuated and only used for service braking.

Types of Questions

ASE certification tests are often thought of as being tricky. They may seem to be tricky if you do not completely understand what is being asked. The following examples will help you recognize certain types of ASE questions and avoid common errors.

Each test is made up of 40 to 80 multiple-choice questions. Multiple-choice questions are an efficient way to test knowledge. To answer them correctly, you must think about each choice as a possibility, and then choose the one that best answers the question. To do this, read each word of the question carefully. Do not assume you know what the question is about until you have finished reading it.

Multiple-Choice Questions

One type of multiple-choice question has three wrong answers and one correct answer. The wrong answers, however, may be almost correct, so be careful not to jump at the first answer that seems to be correct. If all the answers seem to be correct, choose the answer that is the most correct. If you readily know the answer, this kind of question does not present a problem. If you are unsure of the answer, analyze the question and the answers. For example:

Question #1

When considering a school bus service brake relay valve:

    A. air lines connect the delivery ports to the service brake chambers.
    B. when the brakes are released the inlet valve is open.
    C. when the brakes are released the exhaust valve is closed.
    D. the valve is balanced when pressure in the brake chamber equals reservoir pressure.

Analysis:

Answer A is correct. Air lines connect the service brake relay valve delivery ports to the rear axle service brake chambers.
Answer B is wrong. When the brakes are released the inlet valve is closed, not open, in the service brake relay valve.
Answer C is wrong. When the brakes are released the exhaust valve is open-not closed-in the service brake relay valve.
Answer D is wrong. The service brake relay valve is in a balanced position when the air pressure in the rear brake chambers equals signal pressure.

EXCEPT Questions

Another type of question used on ASE tests has answers that are all correct **EXCEPT** one. The correct answer for this type of question is the answer that is wrong. The word **"EXCEPT"** will always be in capital letters. You must identify which of the choices is the wrong answer. If you read quickly through the question, you may overlook what the question is asking and answer the question with the first correct statement. This will make your answer wrong. An example of this type of question and the analysis is as follows:

Question #2

When diagnosing a service brake relay valve, all of the following apply **EXCEPT**:

    A. inlet-valve leakage is tested with the service brakes released.
    B. apply a soap solution to the area around the inlet and exhaust-valve retaining ring to check exhaust-valve leakage.
    C. exhaust-valve leakage is tested with the brakes applied.
    D. the control port on the service-brake relay valve is connected to the supply port on the brake-application valve.

Analysis:

Answer A is wrong. Inlet-valve leakage is tested with the service brakes released.
Answer B is wrong. You do apply a soap solution to the area around the inlet- and exhaust-valve retaining ring to check exhaust-valve leakage.
Answer C is wrong. Exhaust-valve leakage is tested with the brakes applied.
Answer D is correct. This is the **EXCEPT**ion because the control port on the service-brake relay valve is not connected to the delivery port on the brake-application valve.

Technician A, Technician B Questions

The type of question that is most commonly associated with an ASE test is the "Technician A says . . . Technician B says. . . . Who is right?" type. In this type of question, you must identify the correct statement or statements. To answer the question correctly, you must carefully read each technician's statement and judge it on merit to determine if it is true.

Typically, this type of question begins with a statement about some analysis or repair procedure. This is followed by two statements about the cause of the problem, proper inspection, identification, or repair choices. You are asked whether the first statement, the second statement, both statements, or neither statement is correct. Analyzing this type of question is a little easier than the other types because there are only two ideas to consider, although there are still four choices for an answer.

Technician A . . . Technician B questions are really double-true-false questions. The best way to analyze this kind of question is to consider each technician's statement separately. Ask yourself, is A true or false? Is B true or false? Then select your answer from the four choices. An important point to remember is that an ASE Technician A . . . Technician B question will never have Technician A and B directly disagreeing with each other. That is why you must evaluate each statement independently. An example of this type of question and the analysis of it follows.

Question #3

While discussing single- and double-check valves, Technician A says single-check valves are connected between the supply and secondary reservoirs. Technician B says a double-check valve allows air pressure to flow from the lowest of two pressure sources. Who is correct?

    A. A only
    B. B only
    C. Both A and B
    D. Neither A nor B

Analysis:

Answer A is correct. Single-check valves are connected between the supply and secondary reservoirs.
Answer B is wrong. A double-check valve allows air pressure to flow from one of two sources, selecting the higher of the two and routing it to its outlet.
Answer C is wrong.
Answer D is wrong.

Questions with a Figure

About 10 percent of ASE questions will have a figure, as shown in the following example:

Question #4

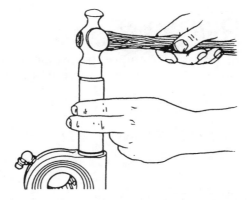

As in the figure, you are installing a seal on a worm gear of a slack adjuster. Technician A says to install the lip of the seal toward the outside of the adjuster. Technician B says not to hit the seal after it bottoms in the bore. Who is right?

    A. A only
    B. B only
    C. Both A and B
    D. Neither A nor B

Analysis:

Answer A is wrong. Answer A is a good choice because it says you install the lip of the seal toward the outside of the adjuster. Yet, it is not the correct answer because both technicians are right.
Answer B is wrong. Answer B is a good choice because you do not have to hit the seal after it bottoms in its bore. Yet, it is not the correct answer because both technicians are right.
Answer C is correct because both technicians are right.
Answer D is wrong.

Most-Likely Questions

Most-Likely questions are somewhat difficult because only one choice is correct, while the other three choices could be nearly correct. An example of a Most-Likely question is as follows:

Question #5

During a brake inspection, a technician tests the air-supply system and finds that the buildup is time slow. Which of the following is the Most-Likely cause?

- A. A clogged compressor inlet filter
- B. A leak in a brake chamber
- C. An air leak in the chassis suspension
- D. A restriction in the governor line

Analysis:

Answer A is correct. A clogged compressor inlet filter is the Most-Likely cause of slow buildup because it stops the compressor cylinder from breathing efficiently.
Answer B is wrong. A leak in a brake chamber is not as likely a cause for slow buildup.
Answer C is wrong. An air leak in the chassis suspension is an unlikely cause.
Answer D is wrong. A restriction in the governor line could more likely cause high system pressure.

LEAST-Likely Questions

For this type of question, look for the choice that would be the LEAST-Likely cause of the described situation. Read the entire question carefully before choosing your answer. An example is as follows:

Question #6

Which of the following is the LEAST-Likely cause of wheel-bearing failure?

- A. Overloading
- B. Contamination
- C. A damaged-race/axle housing
- D. Improper lubricant

Analysis:

Answer A is correct. Overloading is the LEAST-Likely cause of a wheel-bearing failure because they are designed with a significant margin of safety.
Answer B is wrong. Contamination will cause high-heat buildup, destroying the bearing.
Answer C is wrong. Axle housing damage may cause wheel bearings to fail.
Answer D is wrong. Improper or incompatible lubricant will cause failure of the bearing.

Summary

There are no four-part multiple-choice ASE questions having "none of the above" or "all of the above" choices. ASE does not use other types of questions, such as fill-in-the-blank, completion, true-false, word-matching, or essay. ASE does not require you to draw diagrams or sketches. If a formula or chart is required to answer a question, it is provided for you. There are no ASE questions that require you to use a pocket calculator.

Testing Time Length

An ASE test session is four hours and fifteen minutes. You may attempt from one to a maximum of four tests in one session. It is recommended, however, that no more than a total of 225 questions be attempted at any test session. This will allow for just over one minute for each question.

Visitors are not permitted at any time. If you wish to leave the test room, for any reason, you must first ask permission. If you finish your test early and wish to leave, you are permitted to do so only during specified dismissal periods.

Monitor Your Progress

You should monitor your progress and set an arbitrary limit to how much time you will need for each question. This should be based on the number of questions you are attempting. It is suggested that you wear a watch because some facilities may not have a clock visible to all areas of the room.

Registration

Test centers are assigned on a first-come, first-served basis. To register for an ASE certification test, you should enroll at least six weeks before the scheduled test date. This should provide sufficient time to assure you a spot in the test center. It should also give you enough time for study in preparation for the test. Test sessions are offered by ASE twice each year, in May and November, at over six hundred sites across the United States. Some tests that relate to emission testing also are given in August in several states.

To register, contact Automotive Service Excellence/American College Testing at:

ASE/ACT
P.O. Box 4007
Iowa City, IA 52243
Toll Free: 866-427-3273
www.ase.com

4 Overview of the Task List

Brakes (Test S4)

This section includes the task areas and task lists for this test and a written overview of the topics covered in the test.

The Task List describes the actual work you should be able to do as a technician, and that you will be tested on by ASE. This is your key to the test and you should review this section carefully. We have based our sample test and additional questions upon these tasks, and the overview section will also support your understanding of the Task List. ASE advises that the questions on the test may not equal the number of tasks listed; the Task Lists tell you what ASE expects you to know how to do and to be ready to be tested on.

At the end of each question in the Sample Test and Additional Test Questions section, a letter and number will be used as a reference back to this section for additional study. Note the following example: **A1.6.**

Task List

A. Air Brakes Diagnosis and Repair (34 Questions)

1. Air Supply and Service Systems (16 Questions)

Task A1.6 Inspect, test, adjust, or replace system pressure controls (governor/relief valve), unloader assembly valves, pressure protection valves, and filters

Example:

1. Technician A states that the difference between governor cut-in and governor cut-out must not exceed 25 psi. Technician B states that when the difference between governor cut-in and governor cut-out exceeds 25 psi, the governor must be adjusted. Who is right?
   A. A only
   B. B only
   C. Both A and B
   D. Neither A nor B

   (A1.6)

Question #1
Answer A is correct because Federal Motor Vehicle Safety Standards (FMVSS) 121 requires that the difference between governor cut-in and governor cut-out must not exceed 25 psi.

Answer B is wrong because the only adjustment on an air system governor is the governor cut-out value. When the difference between governor cut-in and governor cut-out exceeds 25 psi, the governor is defective and must be replaced.

Answer C is wrong because only one technician is correct.

Answer D is wrong because one technician is right.

Task List and Overview

A. Air Brakes Diagnosis and Repair (34 Questions)

1. Air Supply and Service Systems (16 Questions)

Task A1.1   Diagnose poor stopping, air leaks, pulling, grabbing, or dragging problems caused by supply and service system malfunction; determine needed repairs.

The potential energy of an air brake system is compressed air: the action of the driver's foot on a brake pedal contributes nothing to developing this potential energy, which is produced by an engine-driven air compressor.

A vehicle air brake system can be divided into three circuits: the supply circuit, the control circuit, and the foundation assembly. The supply circuit is responsible for making compressed air at the correct system pressure available to the control circuit. The control circuit manages the service and parking brake functions of the vehicle. The foundation brakes are the mechanical shoes, drums, rotors, and pads that stop the bus and are managed by the control circuit. A failure within any of these three circuits can cause anything from a minor malfunction to a complete brake failure.

School bus brakes are required to be balanced. A balanced braking system can be defined as one in which the braking pressure reaches each actuator at the same moment and at the same pressure level. Factors that affect brake balance are application and release times. To meet balanced performance requirements, vehicle manufacturers match all the brake system valves and components including the hose size and fitting geometry used in the system. Air application and release performance is dependent on both the size and volume of chambers and the distance the air must travel.

An inspection for leaks in an air system is an important part of any brake system inspection. Every component and the entire plumbing circuit of the air brake system should be checked.

Air-related causes of dragging brakes on a vehicle air brake system can include a leaking hold-off diaphragm, low hold-off pressure in the spring brake section of the brake chamber, spring brake control valve problems, low-system pressure causing partial application of the spring brakes, and sticking service application and relay valves.

The majority of braking on a typical highway involves application pressures of 20 psi or less. Full application pressure stops are rare, but the school bus must be capable of them. A school bus air system must be able to accommodate 4–6 full reserve stops with the brakes properly adjusted.

Brakes that are mechanically out of adjustment require a greater volume of application air to effect a stop.

Unlike hydraulic fluid, air is compressible. The larger the volume of air, the higher its compressibility. So brake timing lags are greater in air brake systems than in hydraulic brake systems. When a brake control signal has to travel from the application valve in the tractor to an actuator valve in the bus the lag time depends on the distance the signal has to travel.

To perform a full pressure balance and pressure buildup (timing) test, a technician will need a pair of test hoses and a duplex gauge. Ideal balance timing in a vehicle is defined as each axle receiving identical air pressure simultaneously on a brake application.

Task A1.2 Check air system build-up and recovery time; determine needed repairs.

System air buildup times are defined by federal legislation, specifically FMVSS 121. This legislation defines the required buildup times and values. A common check performed by enforcement agencies requires that the supply circuit on a vehicle be capable of raising air system pressure from 85–100 psi in 25 seconds or less.

Failure to achieve this buildup time indicates a worn compressor, defective compressor unloader assembly, supply circuit leakage, or a defective governor.

Task A1.3 Drain air reservoir tanks; check for oil, water, and foreign material; determine needed repairs.

Air reservoirs store and provide air for the vehicle air braking system. Servicing of the reservoirs consists of inspection, draining the tanks, and performing leakage tests. Air brake systems must have at least three air reservoirs. They may have more. Compressed air from the compressor is delivered to the supply reservoir. The supply reservoir is also known as a wet tank. The supply reservoir feeds air to primary and secondary reservoirs, which supply the brake circuit with air.

A wet tank or supply tank is so named for the moisture that forms in such a tank when hot compressed moisture-laden air cools and condenses on the tank inside walls.

Reservoirs may be equipped with either automatic or manual drain valves. Daily draining of manual drain valves is recommended to keep the system free of contaminants and moisture. Automatic reservoir drain valves should be checked for proper operation, also on a daily basis.

Supply reservoirs are equipped with a safety valve also known as a pop-off valve. This valve is designed to trip at 150 psi. This protects the system in the event of a governor failure.

Air reservoirs on school buses are pressure vessels. They are hydrostatically tested after manufacture. The inside wall is treated with a corrosion-protection coating. Air reservoirs should never be repair-welded, first because they are pressure vessels and second because the internal coating is destroyed. Leaking tanks should be replaced.

Evidence of oil in a wet tank often indicates the compressor is pumping oil through the system. Oil can damage valving throughout the system, so the source of oil contamination must be determined immediately and repaired.

Task A1.4 Inspect, adjust, align, or replace air compressor drive belts, pulleys, tensioners, drive gears, and couplings.

Some compressors, especially those on smaller engines or in those cases where an air brake system has been retrofit, have air compressors driven by the engine using a belt and pulley. Belt sets, pulleys, and idlers should be routinely inspected for indications of wear and axial run out. Belt tension should be set to specification using a belt tension gauge. Cracks and nicks in a drive belt require that it be replaced. Belt tensioners provide a constant load on drive belts to adjust for slack as the belts wear. They should be checked in accordance with OEM specifications.

Task A1.5 Inspect, repair, or replace air compressor, air cleaner, oil and water lines and fittings.

Many compressors are two-cylinder pumps that are balanced units; these do not have to be timed to the engine on installation. Ensure that the oil feed tube is correctly aligned in the compressor crankshaft and that the gear teeth are not damaged.

Other compressors are not self-balanced units, usually single-cylinder types, and these must be timed to the engine on installation. Ensure that the manufacturer's service literature is consulted when installing compressors that must be timed to the engine.

The basic air compressor operates much like an internal combustion engine. It consists of a crankcase, cylinder block, and cylinder head. The crankshaft is supported in the crankcase by main bearings. Connecting rods are connected to throws on the crankshaft at their big end. The piston assemblies are connected to the small end of the crankshaft by means of a wrist pin. As the crankshaft is rotated, the pistons reciprocate in the cylinder bores in the compressor block. Piston rings seal the piston in the cylinder bore and control an oil film on the cylinder wall.

The cylinder head assembly is equipped with discharge valves and discharge ports. It also has reed-type inlet valves. On the piston downstroke, the inlet valve is unseated by lower pressure in the cylinder, and a charge of air is induced into the cylinder. When the piston passes through the bottom of its travel to be driven upward, the inlet valves are seated by the cylinder pressure. As the piston continues to be driven upward, pressure in the cylinder rises, and when it exceeds the pressure at the discharge port, the discharge valve opens and compressed air is forced out of the cylinder into the discharge line.

An unloading mechanism in the cylinder head is controlled by a governor. The governor uses an air signal to cycle the compressor through effective (pumping) and unloaded (not pumping) cycles. The air signal from the governor acts on the unloader assembly which holds the inlet valves open throughout the cycle. When this occurs, the compressor is unable to compress air as the cylinder cannot be sealed; it is therefore unloaded.

The compressor is pressure-lubricated by the engine lubrication system. Pressurized oil is usually delivered to the compressor crankshaft by means of a tube from a bore in the drive gear. The crankshaft main bearings, crankshaft, and piston assemblies require lubrication. Compressors are liquid-cooled by the engine cooling system through inlet and outlet lines located in the cylinder head. The act of compressing air heats it up. Therefore, compressors are required to be cooled to keep air discharge temperatures at 300° F or less.

When troubleshooting a noisy compressor, a technician must determine the source of the noise. Squeaking drive belts or pulleys, failed bearings, and lubrication-related problems are often the cause of a noisy compressor.

An air compressor is the source of compressed air to all the vehicle pneumatic components, including windshield wipers, steering assist units, suspensions, and air starters.

Task A1.6 Inspect, test, adjust, or replace system pressure controls (governor/relief valve), unloader assembly, valves, pressure protection valves, and filters.

Many different air valves are used to control, regulate, or modulate the air in a braking system. These valves guide the direction of flow and control the amount of pressure in the air system.

The governor manages system pressure. It monitors pressure in the supply tank by means of a line directly to it. This pressure acts on a diaphragm and spring within the governor. The spring tension is adjustable. The governor manages compressor-loaded and unloaded cycles. Loaded cycle is the compressor-effective cycle-that is, the compressor is pumping air. Unloaded cycle is when the compressor is being driven by the engine but not actually compressing air; it induces a charge of air through the inlet ports on piston downstroke and forces it out through the same inlet port on the upstroke.

When a compressor is driven by the engine, it is in its loaded cycle until it receives an air signal from the governor to put it into unloaded cycle. This signal acts on the unloader assembly in the compressor cylinder head. The function of the unloader assembly is to hold the inlet valves open, that is, off their seats.

The governor controls the compressor loaded and unloaded cycles. It defines the system pressure. Governed pressure in most air brakes equipped vehicles is set at values between 110 and 130 psi, with 120 psi being typical. Governed pressure is known as cut-out pressure, the pressure at which the governor outputs the unloader signal to the compressor. The unloader signal is maintained until pressure in the supply tank drops to the cut-in value. Cut-in pressure is required by FMVSS 121 to be no more than 25 psi less than that of the cut-out value. The difference on most systems ranges between 20 and 25 psi.

Governor operation can be easily checked. One method is to drop the air pressure in the supply tank to below 60 psi and with the vehicle's engine running, build the pressure. A master gauge should be used to record the cut-out pressure value. This should be exactly at the specification value. If not, remove the dust boot at the top of the governor, release the locknut, and turn the adjusting screw either clockwise or counterclockwise, to lower or raise the cut-out pressure.

Governors seldom fail but they do not have check valves. If the unloader signal is not delivered to the compressor unloader assembly, high system pressures will result. If the safety pop-off valve on the supply tank trips (at 150 psi), this is usually an indication of governor or compressor unloader malfunction.

The only adjustment on an air governor is the cut-out pressure value. If the difference between governor cut-out and cut-in is out of specification, the governor must be replaced.

Task A1.7 Inspect, repair, or replace air system lines, hoses, fittings, and couplings.

Brake hose must be replaced to the original length and specification or brake performance will be compromised.

Sizing of lines on the vehicle will affect both application and release timing of the brakes. Every fitting used in brake system plumbing affects the fluid dynamics and care should be taken to always replace fittings with those similar to the original ones. Replacing a 45° elbow with a 90° elbow is equivalent to adding 7 additional feet of brake hose, and this will affect pneumatic timing.

Brake hose should be securely clamped away from moving components when installed. When reusing dryseal fittings, the nipples and seats should be inspected. Department of Transportation (DOT) approved hose should be used when replacing defective hose.

Brake hose can fail internally, sometimes forming a rubber flap that can act as a check in the line, permitting air to flow toward a valve but trapping it there.

Task A1.8 Inspect, test, clean, or replace air tank relief (pop-off) valves, one-way check valves, drain cocks, spitter valves, heaters, wiring, and connectors.

A system safety valve is usually located in the supply tank. It is designed to trip at a pressure value of 150 psi. It is nonadjustable and consists of a ball seat and spring. Its function is to relieve system air if the pressure builds to a dangerously high level, such as would occur when a governor failed.

The supply tank feeds the primary and secondary reservoirs of the brake system. Each is pressure-protected by means of a one-way check valve. One-way check valve operation can be verified by removing the supply and checking back leakage.

One-way check valves are used variously throughout an air brake system to pressure-protect and isolate portions of the circuit.

Automatic drain plugs can become plugged with sludge (oil and water residues) and may require periodic cleaning. Should oil and excessive water be evident at the drain plugs, check the air dryer and/or compressor.

Task A1.9 Inspect, clean, repair, or replace air dryer systems, filters, valves, heaters, wiring, and connectors.

Moisture in an air system can be very damaging. Ambient moisture is a problem any time the relative humidity is high, so it is both a summer and a winter problem. The airborne moisture condenses in the reservoirs as the compressed air cools. Most current systems use air dryers to remove moisture from the compressed air before it gets to the supply tank. These use two principles to remove moisture from the compressed air. The first type is the desiccant type. During the charge cycle, hot compressed air passes through a desiccant pack that adheres the moisture; dry air exits to the discharge port and the supply reservoir. At governor cut-out, an air signal from the governor begins the purge cycle of the air dryer. This acts on the purge piston, which exhausts any water collected in the purge orifice to atmosphere. The second type is the heat exchanger type. The heat exchanger type of air dryer attempts to cool the compressed air to the point that the moisture is condensed; once condensed, it can be separated and dumped by means of a purge valve. Some air dryers use a combination of both principles.

Some air dryers use a heater to prevent icing in cold weather; the heater is thermostat regulated. In cold weather, the thermostat controls the heater cycles to maintain a temperature exceeding 45° F.

Oil can destroy the desiccant pack in an air dryer. When an air compressor fails and pumps its lubricating oil through the system, the desiccant pack becomes contaminated and requires replacement.

Task A1.10 — Inspect, test, adjust, repair, or replace brake application foot/treadle valve, fittings, and mounts; check and adjust brake pedal free play.

The service application valve in a school bus brake system is a floor-mounted foot valve, known as a treadle valve. The treadle valve is actually two valves in one. The upper portion of the valve is the primary section and the lower portion is the secondary or relay section. Each section has a dedicated feed and its own exhaust port. The upper or primary section of the treadle valve is supplied directly from the primary reservoir. The lower or relay section of the valve is supplied by the secondary reservoir.

The treadle valve is actuated mechanically, that is, by foot pressure from the driver. When the driver's foot acts on the treadle valve, the primary piston is forced downward. This movement first closes the primary exhaust port and then modulates air proportional to piston travel to:

1. actuate whatever brakes are plumbed into the primary circuit, the drive axle on a typical bus.
2. actuate the relay or secondary piston, located below the primary section in a dual-circuit application valve.
3. act against the mechanical pressure (foot pressure) to provide brake feel.

The secondary or relay section of the dual-circuit application valve is actuated pneumatically by primary circuit air. This section operates similarly to a relay valve in that a signal pressure value (the air from the primary section) is used to displace a relay piston that then modulates secondary circuit air to whatever brakes/valves are located in the secondary circuit. In a typical school bus air brake system, this would normally be the front axle brakes. Like a relay valve, secondary section is designed to modulate an air pressure value to the secondary circuit, identical to the signal pressure.

Each portion of the valve has its own exhaust port. Primary and secondary circuit air never come into contact with each other in the treadle valve.

In an emergency application of the treadle valve, both the primary and secondary inlet valves are held open and full reservoir pressure is applied to each of the two circuits.

In the event of a total primary circuit failure, no air is available to the primary section of the treadle valve. The dash low pressure warning alerts would trip, and when the treadle valve is depressed, foot pressure would drive the primary piston downward until it mechanically contacts the relay piston to actuate the secondary circuit. There would be zero brake "feel," and there would be greater pedal travel.

In the event of a total secondary circuit failure, the primary section of the treadle valve would function normally. However, the dash low air pressure alert would trip, and the vehicle would have to be brought to a halt using primary circuit source air only.

**Task
A1.11**

**Inspect, test, clean, or replace two-way (double) check valves
and anti-compounding valves.**

Two-way (double) check valves play an important role as a safeguard in dual circuit air brake systems. The typical two-way check valve is a T with two inlets and a single outlet. It outputs the larger of the two source pressures to the outlet port and checks the lower value source; the valve will shuttle in the event of a change in the source pressure value. In other words, it will always prioritize the higher source pressure. Two-way check valves provide a means of providing the primary circuit with secondary circuit air and vice versa in the event of a circuit failure. In the event that both source pressures are equal, as would be the case in a properly functioning dual air brake circuit, the valve will prioritize the first source to act on it.

Compounding occurs when a brake foundation is subjected to both mechanical and pneumatic force. A spring brake chamber in park mode has the air acting on the hold-off chamber exhausted from it; this enables the mechanical force of the spring in the chamber to act on the slack adjuster to apply the brakes. As it takes an air pressure of approximately 60 psi to cage the spring brake, it is capable of applying approximately the equivalent of that amount of force to the foundation brakes when no air is acting on the hold-off diaphragm. If, when in this parked condition, a driver made a full service application of the brakes, the mechanical force that the spring was applying to the foundation brakes would be compounded by a further 120 psi acting on the service diaphragm. The result would be a total application pressure 50 percent greater than the specified maximum. To prevent this from happening, anti-compounding valves are used. These operate by dividing service application air between the service and hold-off chambers whenever a service application is made in the absence of air in the hold-off chamber.

Anti-compounding valve operation can be easily verified with a pair of air pressure gauges fitted to the service and hold-off lines while the vehicle is parked.

Failure of an anti-compounding valve can result in twisted S-cam shafts, spline damage, and slack adjuster damage.

**Task
A1.12**

**Inspect, test, repair, or replace stop and parking brake light circuit
switches, wiring, and connectors.**

Functioning brake lights are required in all highway vehicles. School buses with air brakes require two brake light switches, either of which can illuminate the vehicle brake light circuit.

The service stop light switch is a normally open, air-actuated switch, plumbed into the service brake circuit. The switch is closed by a small application pressure acting on it.

The parking brake light switch works oppositely. This switch is designed to close the electrical circuit when no air is acting on it and to open when air is charged to the hold-off chambers.

The operation of both switches can be verified with a digital multimeter (DMM). Voltmeter mode should be used when testing either switch in circuit and the ohmmeter, when out of circuit.

**Task
A1.13**

**Inspect, test, repair, or replace brake relay valve, quick-release valves,
and limiting quick-release valves.**

Relay valves permit a remote air signal (from the treadle valve) to effect service braking with an air supply close to the brake chambers. The valve is controlled by a signal that is plumbed to its service port. At the supply port, system air pressure is available to the valve from a close-by air reservoir. When the relay valve receives an air signal, this pressure acts on the relay piston. When the relay piston is displaced, it modulates the air available at its supply port, to actuate the service brake chambers connected to its delivery ports. One relay valve can typically manage two service brake chambers. The valve is usually designed so that the signal value is identical to the output value modulated to the delivery ports. When the signal pressure is dropped or relieved, the relay valve exhausts the service supply air returning from the service chambers, and a retraction spring returns the relay valve to its neutral position.

A sticking relay piston in the relay valve will cause supply air to be modulated to the service chambers when there is no signal pressure; this causes dragging brakes or locked brakes depending on the location of the relay piston. Moisture in the air system may freeze and prevent the relay retraction spring from returning the relay piston.

Relay valve operation can be verified using a pair of gauges that monitor signal pressure and delivery.

Quick-release valves are mounted close to the brake chambers or components they serve. They are used throughout the air brake system and other vehicle air systems to speed release times.

The typical quick-release valve has a single inlet port and two outlet ports. When air is charged to the inlet port, the valve acts like a T fitting and simply divides the air to the two outlets. However, when the air pressure supplied to the inlet is exhausted at the application or control valve, the air being outputted from the quick release to the brake chambers is exhausted at the quick-release valve.

Because of its simplicity, quick-release valves seldom fail. They are vulnerable to external damage because of their exposed location and contaminated air, which may cause exhaust port leakage.

Task A1.14

Inspect, test, and replace inversion/emergency (spring) brake control valve(s).

The spring brake valve supplies air to the hold-off chambers to release the spring brakes and enable the vehicle to move. During normal operation, it limits the hold-off pressure to the spring brake chambers to a value of around 90 psi; this speeds application times in the event of an emergency and permits a consistent hold-off pressure (system pressure fluctuates between cut-in and cut-out pressure values).

Spring brake control valves are often incorporated in multifunction valves that contain parking, service, and inversion functions.

Task A1.15

Inspect, test, repair, or replace low-pressure warning devices.

FMVSS 121 requires that a driver receive a visible alert when the system pressure drops below 60 psi; in most cases this is accompanied by an audible alert, usually a buzzer. A low air pressure warning device is fitted to both the primary and secondary circuits. This is a simple electrical switch that can be plumbed anywhere into a system requiring monitoring. The switch is electrically closed whenever the air pressure being monitored is below 60 psi; when the air pressure value exceeds 60 psi, the switch opens.

Verifying the operation of a low air pressure warning switch can be done by pumping the service application valve until the system pressure drops to the trigger value.

Both the primary and secondary circuit air pressure must be monitored by a dash-located gauge. The required visible warning is usually a dash warning light.

Task A1.16

Inspect, test, and replace air pressure gauges, lines, and fittings.

Air pressure gauge operation can be verified by using a master gauge, a good quality, liquid-filled gauge that uses a Bourdon principle of operation. When troubleshooting vehicle air pressure management problems, the vehicle gauges should not be relied on. Evidence of oil behind the glass in an air pressure gauge is an indication of a compressor pumping oil.

Task A1.17

Install air dryer system, hoses, fittings, and wiring.

An air dryer is installed downstream from the compressor and upstream from the supply (wet) tank. The OEM recommendations for the compressor to dryer hose size and material must be adhered to. The dryer should be securely clamped to a chassis frame rail. The hose pipe that connects the compressor to the dryer is designed to dissipate heat and may be manufactured from braided steel or copper pipe. The air dryer purge valve is actuated by the governor unloader signal, so a connection must be made between the two, usually with $\%$ pneumatic hose. Many current air dryers are heated to prevent moisture from freezing the purge valve; the heater unit is thermostat regulated. The dryer heater is key-on energized, protected by an 8-amp fuse and requires the use of at least 16-gauge connection wire. To test electrical operation, use a DMM remembering that the heater core will only activate when the thermostat senses temperatures below 40° F.

Task
A1.18

Perform antilock brake system (ABS) warning lamp start-up test; determine needed repairs.

ABS systems perform a self-test on start-up. This test consists of sequentially actuating the solenoids which makes an audible clicking noise. If a circuit malfunction is read by the ABS module, this can be read by interpreting flash codes. It should be noted that only electronic and electrical problems can be read by the ABS module. This occurs by checking the input and output circuits of the ABS management system. Always observe the OEM recommended troubleshooting procedure and remember that, unless otherwise indicated, a DMM must be used when performing system checks.

Task
A1.19

Diagnose poor stopping, lock-up, pulsation, and noise problems on antilock brake systems (ABS); determine needed repairs.

It is a requirement of current ABS that the system is engineered with full redundancy. This means in the event of a sub-circuit or full circuit failure, braking reverts to full pneumatic brake operation. ABS is simple to troubleshoot. In most cases, when there is a poor braking complaint and no evidence of an ABS fault code, the problem source is outside of the ABS electronic circuit. Note that pneumatic ABS cycle frequencies are much slower than their hydraulic equivalents and, as a consequence, not as smooth. Maximum cycle frequencies in current school bus ABS are around 7 times per second versus about 3 times that frequency (20 times per second) for hydraulic brakes. This lower frequency operation can produce a pulsing sensation when the brakes are functioning properly and should be of concern only if they begin to produce a snub effect.

Task
A1.20

Test, adjust, or replace antilock brake system (ABS) wheel-speed sensors and tone/exciter rings/tooth.

Wheel-speed sensors produce more than their share of ABS complaints largely because of their vulnerable location. Gapping complaints will usually be identified by the system self-diagnostics and are easy to remedy. Wheel-speed sensors function using inductive pulse generation. A reluctor (known as a pulse generator, tone wheel, exciter ring or chopper wheel) cuts through a magnetic field, inducing an AC current which acts as a signal to the ABS module. The module receives the AC electrical signal and correlates its frequency to a wheel speed value.

Wheel speed sensors should always be checked using the OEM recommended procedure; this can be done by using a DMM in resistance mode or by a dynamic check using V-AC mode.

2. Mechanical/Foundation (12 Questions)

Task
A2.1

Diagnose poor stopping, brake noise, pulling, grabbing, or dragging complaints caused by foundation brake, slack adjuster, and brake chamber problems; determine needed repairs.

The S-cam, wedge, or disc brake mechanism itself, linings or pads, and related parts such as brake chamber(s), slack adjuster(s), and parking brake components generally make up what is called the foundation brake assembly.

All vehicle braking requires that kinetic energy (the energy of motion) be converted to heat energy by means of friction; the heat energy must then be dissipated to atmosphere. The foundation brake assembly consists of those brake components that are responsible for effecting the retarding effort required to stop a vehicle.

In a typical S-cam, shoe/drum assembly, slack adjusters connect the brake actuation chambers with the foundation brake assembly. Slack adjusters are levers so their length is critical. The greater the distance between the centerline of the S-cam and the point at which the brake chamber clevis connects with the slack adjuster, the greater its leverage. Slack adjusters also convert the linear force produced by the brake chamber into torque (twisting force). Current slack adjusters are required to be automatically adjusting.

The geometry of the relationship between the brake chamber and slack adjuster requires that the angle between the slack adjuster and the chamber pushrod be 90 degrees when the brake is in the fully applied position. In any position other than 90 degrees, the slack adjuster has less mechanical advantage. As brake shoe friction surfaces wear, this 90-degree angle must be maintained by periodic adjustment of manual slack adjusters or functioning automatic slack adjusters.

Both parking and service brake performance relate directly to brake adjustment.

Foundation brake problems can cause unbalanced braking, grabbing, not releasing, or failure to apply.

Some causes of foundation brake problems are: use of improper replacement components, cargo weight overloads, lining contamination, poorly maintained or installed brakes, broken or malfunctioning brake components, brakes out of adjustment, and overheating.

Task A2.2

Inspect, test, adjust, repair, or replace service brake chambers, diaphragm, clamp, spring, pushrod, clevis, and mounting brackets.

To service a spring brake chamber, it is necessary to "cage" the main spring before any service is performed. A spring brake is normally manually caged by tightening a nut on a supplied release tool. The release tool has a pair of lugs that engage to an internal cage plate; when the cage nut is tightened, the cage plate compresses the main spring.

If a clevis is not positioned correctly onto the pushrod, the brake may not completely release. A clevis has a threaded yoke bore for a pushrod to engage. Most manufacturers provide slack adjuster installation templates facilitating correct installation.

A 0.060-inch clearance between the clevis pin and the clevis collar is considered the maximum allowable before replacement. Replace the clevis if the collar hole is worn beyond this limit. An automatic slack adjuster automatically adjusts the clearance between the brake linings and the brake drum or rotor. The slack adjuster controls clearance by sensing stroke length for the air brake chamber, so no manual adjustment is required.

Common causes of sluggish service braking in an air system are problems concerning foundation brake geometry. Problems such as an obstruction in a brake chamber or poor alignment of the brake linkage can produce apply- and release-rate performance problems.

Spring brake chamber assemblies should not be repaired but replaced as an assembly, after they have been disarmed.

When the park control valve closes, air in the hold-off section of the spring chamber is exhausted, allowing the main spring to apply the parking brake.

Task A2.3

Inspect, tests, adjust, repair, or replace manual and automatic slack adjusters.

The slack adjuster is critical in maintaining the required free play and actuation angle. Current slack adjusters are required to be automatically adjusting, but their operation must be verified routinely.

Slack adjusters connect the brake chambers with the foundation assemblies on each wheel. They are connected to the pushrod of the brake chamber by means of a clevis yoke (threaded to the pushrod) and pin. Slack adjusters are spline-mounted to the S-cams and positioned by shims and an external snap ring.

The slack adjuster converts the linear force of the brake chamber rod into rotary force or torque and multiplies it. The distance between the slack adjuster S-cam axis and the clevis pin axis will define the leverage factor; the greater this distance, the greater the leverage.

The objective of brake adjustment, whether manual or automatic, is to maintain a specified drum to lining clearance and a specified amount of free play. Free play is the amount of slack adjuster stroke that occurs before the linings contact the drum.

Slack adjusters are lubricated by grease or automatic lubing systems. The seals in slack adjuster are always installed with the lip angle facing outward; when grease is pumped into the slack adjuster, grease should exit the seal lip when the internal lubrication circuit has been charged.

Automatic slack adjusters may require periodic adjustment. The method varies by manufacturer. Manual slack adjusters are adjusted by rotating the adjusting screw until the shoes are forced into the drums, then backing off the adjusting screw to set the specified free play. When a wheel-up adjustment is performed, free play is set to minimum drag of the shoe/drum relationship. Automatic slack adjuster free stroke should typically measure 0.75" and applied stroke 1.25" when functioning properly.

When replacing a failed slack adjuster, the distance between the axis of the S-cam bore and the clevis pin bore must be maintained; a difference of one-half inch can greatly alter the brake torque and unbalance the brakes.

Task A2.4
Inspect or replace cams, rollers, shafts, bushings, seals, spacers, and retainers.

When performing foundation brake servicing, all of the hardware mounted on the axle spider should be inspected and replaced if required. There should be no radial play of the S-camshaft, rollers and cams should be inspected for flat spots, and it is good practice to replace any spring steel components such as retraction springs and snap rings.

Spider-fastener integrity should be checked at each brake job. Spiders should be inspected for cracks at each brake job.

S-cam bushing seals should be installed so that the spider external bushing sealing lip faces inboard and the internal bushing sealing lip faces outboard from the bushing; this allows excess grease to exit from the inboard side of the bushing and prevents it from being pumped into the foundation assembly.

S-cam profiles depend on friction and should never be lubricated.

Task A2.5
Inspect, or replace brake spider, shields, anchor pins, bushings, and springs.

In an S-cam, air brake system, the foundation assembly components are mounted to a spider assembly bolted to the axle. The brake shoes are mounted to the spider by means of anchors on which they pivot. An S-cam is responsible for spreading the shoes and forcing them against the drum; the S-cam profile acts on rollers clipped to the pallet end of the shoe assembly. When brake torque is applied to the S-camshaft by brake chamber linear force acting on a slack adjuster (lever), the cam-profiles act on the rollers to force the shoes into the brake drum.

When foundation brakes are serviced, the friction facings (linings) or the shoes assemblies themselves are replaced. It is good practice to replace the anchor pins, bushings, and retaining springs during a brake job. Brake spiders will generally last the life of the vehicle. When removing anchor pins, use a puller rather than sledgehammer or impact tools that can damage the anchor eyes and bend the spider assembly. Braking is effected by friction, so be careful as to how grease and oil are used around brake components. Generally bushings require grease, but avoid applying grease to S-cam profiles and actuator rollers, which require friction to function properly.

Task A2.6

Inspect, clean, rebuild or replace, and adjust air disc brake caliper assemblies.

Air-actuated school bus disc brakes have some similarities to hydraulically actuated disc brakes. Air disc brakes use an air chamber and pushrod to apply braking torque to a powershaft. The powershaft has an external helical gear that acts on an actuator nut to create the clamping force required by the caliper to effect retarding effort on a rotor that turns with the wheel assembly. In other words, axial force applied to the powershaft is converted to clamping force by the caliper assembly.

Within the caliper assembly, pistons transmit the clamping force to brake pads which effect retarding effort by converting the kinetic energy (energy of motion) to friction and then heat.

Short outboard friction pad service life is usually an indication of caliper assembly seizure, that is, the sliding action of the brake caliper ceases. The caliper slide pins are usually at fault. The rotors used on bus air disc brakes are usually vented to aid in the dissipation of brake heat.

Task A2.7

Inspect and replace brake shoes, linings, or pads.

Bus air brake shoes are, in most cases, fixed anchor assemblies mounted to the axle spider and actuated by an S-camshaft. Brakes can have their friction facings or linings mounted to the shoe by bonding, riveting, or by fasteners. In most current applications, the shoes are remanufactured and replaced as an assembly, that is, with the new friction facing already installed. When reusing shoes, they must be checked for arc deformities usually caused by prolonged operation with out-of-adjustment brakes. The lining blocks are tapered and seldom require machine arcing.

The brake shoes must be fitted with the correct linings. Lining requirements can change with different vehicles and different applications. When reconditioning shoes with bolted linings, if the original fasteners are reused, the lock washers should at least be replaced. Riveted linings should be riveted in the manufacturer's recommended sequence.

The friction rating of linings is coded by letter codes. Combination lining sets of shoes are occasionally used; these use different friction ratings on the primary and secondary shoes. When combination friction lining sets are used, care should be taken to install the lining blocks in the correct locations on the brake shoes.

It is good practice to replace the brake linings on all four wheels of a school bus on a preventive maintenance inspection (PMI) schedule. When the linings on a single wheel are damaged ,such as in the event of wet seal axle lube failure, the linings of both wheel assemblies on the axle should be replaced to maintain brake balance.

The friction pads in air disc brake assemblies should also be changed in paired sets, that is, both wheels on an axle. Vented disc brakes may have thicker inner pads than outer pads, while solid disc brakes usually have equal thickness in the inner and outer pads. Heat transfer is generally more uniform in the solid disc assemblies than vented ones, requiring thicker inboard pads.

Task A2.8

Inspect, resurface, or replace brake drums or rotors.

If brake drums are to be reused (recent practice and the low cost of drums often results in drum replacement with a brake job), they should be inspected for size specification, heat discoloration, scoring, glazing, threading, concaving convexing, bellmouthing, and heat checking. Used brake drums can seldom be successfully machined due to heat tempering. It is good practice to machine new drums before installation because of warpage caused by incorrect storage practices after manufacture.

A drum micrometer is used to check brake drum diameter. Ensure that manufacturer machining and discard limits are strictly observed.

Brake rotors must be visually inspected for heat checking, scoring, and cracks. Rotors must be measured for thickness with a micrometer. A dial indicator is used to check rotor run out and parallelism. In most applications, rotors tend to outlast drums and are often reused after a brake job. They must be turned within legal service specifications using a heavy-duty rotor lathe. Ensure that manufacturer machining and discard limits are strictly observed.

3. Parking Brakes (6 Questions)

Task A3.1

Inspect or replace parking (spring) brake chamber.

Federal legislation requires that all vehicles with air brakes be equipped with a mechanical parking and emergency brake system. The parking brake system is generally a separate system from the service brake system, with its own lines, chambers, and valves. The mechanical braking force is obtained by springs located in spring brake chambers. These use air acting on a hold-off diaphragm to release the spring brake.

The spring in a spring brake chamber requires approximately 60 psi acting on its hold-off chamber to fully release it. It is therefore capable of applying the equivalent of a 60 psi air application when there is no air acting on the hold-off chamber.

Spring brake chambers are dual chamber assemblies that combine a service application chamber in the front section and a spring chamber in their rear section. Their size rating (such as 30/30, 24/24, 30/24) refer to the surface area of the service and hold-off diaphragms. The caging port at the rear of the spring diaphragm is sealed with a plastic seal; if this seal is not in place, the internal components can rapidly corrode.

Spring brake chamber assemblies fail when either the service or hold-off diaphragms puncture or when the spring breaks. Spring breakage is often indicated by an air leak at the hold-off diaphragm that occurs when the fractured spring punctures it. It is essential that spring brakes be handled with caution, even when it is known that the main spring has fractured.

Task A3.2

Inspect, test, or replace parking (spring) brake check valves, lines, hoses, and fittings.

The supply of air to the parking brake circuit is pressure-protected so that in the event of a complete primary or secondary circuit failure, there is sufficient air for a complete apply/release sequence. Restrictions in the lines supplying the hold-off chambers can slow both application and release times. Spring brake valve operation can be verified with an air pressure gauge. Lines supplying the hold-off chambers on each wheel should be of equal length; release timing is unbalanced if this is not the case.

Task A3.3

Inspect, test, or replace parking (spring) brake application and release valve.

Spring brake valves are often incorporated in multifunction valves. It is critically important that when valves malfunction, they be replaced by matching the part numbers, remembering that two valves that appear physically identical may have widely different performance characteristics.

The parking brake system on a school bus manages hold-off pressure delivered to the spring brake chambers. School buses equipped with spring brake chambers on the rear axle manage the hold-off and park cycles with a single dash valve, usually yellow with a diamond/square shape. It is pushed inward to release the parking brakes. When pulled out, it exhausts the hold-off air in the spring brake chambers, putting the vehicle into park/emergency mode.

Task A3.4

Manually release and cage parking (spring) brakes.

Spring brake chambers are usually equipped with a cage bolt. The cage bolt is threaded and has a pair of lugs that engage to an internal cage plate. When a nut is turned down the cage bolt, the cage plate is pulled inwards, compressing the main spring of the spring brake chamber. This releases the parking brake. If the vehicle on-board air supply is available, a spring brake may be caged by chocking the wheels and releasing the parking brakes (supplying air to the hold-off chambers) and then installing the cage bolt to the cage plate with the main spring already compressed. It is important to ensure that the lugs in the cage bolt are properly engaged in the cage plate. Cage plates are manufactured out of aluminum alloy and those in older spring brake assemblies were susceptible to corrosion. Caged spring brake assemblies should always be handled with a great amount of care. Spring brake chambers should always be removed and installed fully caged.

In the past, failed hold-off diaphragms were routinely replaced. This is not current practice and when spring brake hold-off diaphragms fail, a piggy-back assembly (spring brake assembly minus the service chamber) or entire spring chamber is used to repair the condition. The spring brake chamber band clamps should never be removed with the main spring in uncaged condition. Many current spring brake assemblies have chamber clamps that cannot be unbolted.

It is illegal to discard spring brakes without first disarming them. Disarming a spring brake means releasing the main spring. This is achieved by placing the entire spring brake assembly in a disarmament chamber with the main spring uncaged, torch-cutting the spring brake chamber clamps, and separating the chamber. The entire assembly should remain in the disarmament chamber until the main spring can be observed to be free.

B. Hydraulic Brakes Diagnosis and Repair (21 Questions)

1. Hydraulic System (11 Questions)

Task B1.1

Diagnose poor stopping, pulling, dragging, or brake feel complaints caused by hydraulic system problems; determine needed repairs.

The potential energy of a hydraulic brake system is mechanical force created by the action of a driver's foot acting on a brake pedal, usually assisted proportionally by pedal geometry leverage and a power assist system. In any hydraulic circuit, it can be assumed that the hydraulic medium is not compressible. If force is mechanically applied to a liquid in a closed system, it will be transmitted equally by the liquid to all parts of the system. Force applied by a master cylinder is transmitted equally throughout the hydraulic system though it may be modulated by valves in parts of that system.

The hydraulic circuit of a hydraulic brake system consists of a master cylinder, proportioning valves, metering valves, pressure differential valve and wheel cylinders. All school bus hydraulic brake systems have dual circuits, meaning that in the event of a failure in one of the circuits, the other will back it up to effect at least one stop. As in bus air brake systems, the circuits are defined as the primary and secondary circuits. A failure of any moving part within the hydraulic circuit may cause pressure to become entrapped in a portion of the circuit, and this may cause dragging brakes or slow release times.

Task B1.2

Inspect hydraulic system for leaks.

Pressure values within the hydraulic circuit may be tested with pressure gauges. The hydraulic system circuit can be pressurized simply by starting the vehicle engine and applying the brakes by foot pressure. External leaks may be verified by cleaning the externally visible portions of the circuit and applying the brakes. Internal leaks are more difficult to locate. Internal leakage within a master cylinder can be verified by using gauges plumbed to each portion of the hydraulic circuit.

Task B1.3

Check and adjust brake pedal free play.

Most school bus hydraulic brake systems use the brake pedal assembly to provide added leverage to the mechanical force provided by the driver's foot pressure. Brake pedal pushrod adjustment should always be made according to manufacturer's specifications.

Task B1.4

Inspect, test, or replace master cylinder.

The master cylinder converts the mechanical force applied to it by driver foot pressure and the brake booster system into hydraulic pressure to actuate the primary and secondary circuits of the brake system. It usually consists of integral reservoirs (one for each circuit), cylinder housing, compensating ports, return springs, and primary and secondary pistons. The primary piston is actuated mechanically. Pressure developed in the primary portion of the master cylinder charges both the primary circuit and the secondary piston. When the master cylinder is operating normally, the secondary piston is actuated hydraulically, by whatever pressure value is developed in the primary portion of the cylinder, to charge the secondary circuit.

When the mechanical force applied to the primary piston is relieved, return springs acting on both the primary and secondary pistons return them to their original positions, permitting the fluid applied to each circuit to return to the reservoirs. Both sections of the master cylinder are aspirated with brake fluid by fill and compensating ports. Each piston is sealed in its bore by rubber seals. Primary and secondary circuit fluid do not come into contact with each other under normal operation.

When a failure occurs in either circuit, the pressure differential light in the dash will illuminate the first time the brakes are applied following the failure. If the failure occurs in the primary circuit, the primary piston will be forced through its travel without generating any fluid pressure until it contacts the secondary piston and mechanically actuates the secondary circuit. If the failure has occurred in the secondary circuit, the primary circuit will function normally, but the actuation of the secondary piston will result in no pressure delivered to the secondary circuit. In either case, the vehicle should be brought to an immediate standstill and not operated until a repair has been undertaken.

When testing a master cylinder in a hydraulic braking system, a liquid-filled hydraulic test gauge should be used. Deteriorated fluid, deteriorated seals, or a mixture of incompatible fluids may cause sludge and particulate that can plug fill and compensating ports, resulting in slow application times, slow release times, and brake failure.

Master cylinders in school bus hydraulic systems are routinely reconditioned when seals and springs are replaced, cylinders honed, etc. When cleaning master cylinders before reassembly, only isopropyl alcohol should be used to clean components; use of solvents can swell the seals and leave behind corrosive residues.

Task B1.5

Inspect, test, or replace brake lines, flexible hoses, and fittings.

Steel tubing should be checked for wear, dents, kinks, and corrosion. Preflared and preformed tubing helps reduce custom cuts, bending, and flaring of new tubing. Two types of flaring styles and seats are used: International Standards Organization (ISO) and double flare. ISO uses an outward flare, while a double flare creates a double wall at the nipple seat for greater strength. When cleaning brake tubing, only isopropyl alcohol should be used because of its ability to evaporate rapidly and residue-free drying. Residues remain when using other cleaning agents such as soap and water, mineral spirits, and hydraulic brake fluid. When bending tubing, a tube bender should be used and the line filled with fine silica sand to prevent collapsing. Remember, kinked lines create flow restrictions that can affect brake performance.

**Task
B1.6**

Inspect, test, and replace metering (hold-off), proportioning, and combination valves.

A metering valve is used on vehicles equipped with front disc and rear drum brakes. It is required to achieve brake timing balance during light brake applications by withholding the delivery of application pressure to the front disc brakes until pressure exceeds a predetermined value in the circuit responsible for actuating the rear brakes. This lag or delay is required so that hydraulic pressure builds sufficiently in the rear brake hydraulic circuit to overcome the tension of the rear brake shoe return springs and the free travel of the shoes. The objective is to enable simultaneous application of both front and rear brakes. For this reason, the metering valve is sometimes known as a hold-off valve.

When a pressure bleeder is used to bleed any system equipped with a metering valve, the manufacturer's instructions as to how to open the valve must be observed. When manually bleeding a brake system, application of the brake pedal develops sufficient pressure to overcome the metering valve opening pressure.

A proportioning valve is also used on systems combining front disc and rear drums. The proportioning valve is installed in the circuit supplying the rear brakes. Its function is to reduce the application pressure to the rear wheel cylinders and prevent rear wheel lock-up. Disc brakes require higher hydraulic application pressures.

The proportioning valve operation should be verified at each brake inspection or if reported for a rear wheel lock-up condition. To check valve operation, hydraulic gauges should be installed ahead and behind the valve or alternatively, to each of the two circuits.

**Task
B1.7**

Inspect, test, or replace brake pressure differential valve and warning light circuit switch, bulbs, wiring, and connectors.

The pressure differential valve is also known as a brake light warning valve and dash lamp valve. It consists of a cylinder through which primary and secondary hydraulic pressure act on either side of a spool. When the pressure in both the primary and secondary circuits is equal, the spool floats in a neutral position. Should a pressure imbalance occur in either the primary or secondary circuit, the spool will shuttle to one side of the cylinder and in doing so, ground an electrical signal, illuminating a dash warning light.

The pressure differential valve will normally re-center automatically upon the first application of the brakes after repairs are completed. Some pressure differential valves will require manual resetting.

The proportioning valve should be inspected whenever the brakes are serviced. This inspection should include the electrical warning light circuit.

**Task
B1.8**

Inspect, clean, and rebuild or replace wheel cylinders.

A wheel cylinder is the actuator of a hydraulic brake system; it is also known as a slave cylinder. It is supplied with hydraulic brake fluid supplied by the master cylinder, and converts hydraulic pressure to mechanical force at the foundation brake assembly. Wheel cylinders are usually constructed of cast iron for higher durability and lower manufacturing costs.

Most wheel cylinders are double acting. They are machined out of cast iron and house two pistons within a cylinder bore. The pistons are sealed in the cylinder by rubber seals. When the pistons are subjected to hydraulic pressure, they are forced outward to actuate brake shoes, forcing them into the drum. Manual and automatic adjusting mechanisms are integral in the wheel cylinder assembly.

Wheel cylinders are commonly reconditioned. The cylinder bores can be honed and most of the internal components replaced if required. They are vulnerable to contaminates in the hydraulic fluid. When the cylinder bores become scored, the result is fluid leakage from the cylinder past the seals. Seals may fail if exposed to chemical contaminants. Each wheel cylinder is equipped with a bleed port. You crack open this bleed port to purge air from the hydraulic circuit in the bleeding process.

Task B1.9

Inspect, clean, and rebuild or replace disc brake caliper assemblies.

Fixed and sliding or floating calipers are used in hydraulic brake systems. In a fixed-type caliper, the rotor sits in between two or four pistons, and the clamping action of opposed pistons acting on the rotor is responsible for retarding the rotor. In a floating or sliding caliper, one or more pistons sit on one side of the rotor, and the caliper housing is designed to slide when the piston is subjected to hydraulic pressure; this action effects the clamping action of the rotor. The friction surfaces that contact the rotor are brake pads. Disc brakes are not self-energizing and in general greater hydraulic pressures are required to apply them. They operate at higher mechanical efficiencies than drum brakes. Disc brake calipers are reconditioned in the same manner as wheel cylinders. Manufacturer's tolerance specifications must be observed during reconditioning.

Task B1.10

Bleed and/or flush hydraulic system.

It is good practice to replace the system brake fluid at each major brake overhaul, especially if the failure has been caused by seal failure or fluid contamination. Using the manufacturer's specified brake fluid is important to ensure proper operation of the system. The approved heavy duty brake fluid should retain the proper consistency at all operating temperatures. It will not damage rubber cups and helps to protect the metal parts of the brake system against failure. Water, mineral spirits, and gasoline should never be used to flush a hydraulic braking system because of incompatibility with other materials and corrosive effects. Alcohol or compatible brake fluid is always used to flush hydraulic braking systems, and these fluids should not be reused after the flushing is complete.

A container storing brake fluid must always be tightly sealed when not in use to prevent moisture from being absorbed into it. Mineral oil, alcohol, antifreeze, cleaning solvents, and water, even in very small quantities, will contaminate most brake fluids. Brake fluid has a shelf life of 1 year. This shelf life diminishes if the storage container is not completely full. Mixing of different brake fluid can cause coagulation and result in hydraulic failure; it is important to ensure that fluids are compatible when topping up system reservoirs.

Brake systems may be bled using a bleeder ball or manually. The first method is preferred as the operation can be performed by one person. In all cases, system air is purged from the valves and wheel cylinders, in a sequence defined by the manufacturer

Task B1.11

Perform antilock brake system (ABS) warning lamp start-up test; determine needed repairs.

Hydraulic ABS systems also perform a self-test on startup. This test checks the input circuit components (wheel speed sensors), control module processing, and the output circuit (modulator). If a circuit malfunction is read by the ABS module, this can be read by interpreting flash codes. It should be noted that only electronic and electrical problems are be read by the ABS module. Always observe the OEM recommended troubleshooting procedure, and remember that unless otherwise indicated, a DMM must be used when performing system checks.

Task B1.12

Diagnose poor stopping, lock-up, pulsation, and noise problems on antilock brake systems (ABS); determine needed repairs.

It is a requirement of current ABS that the system is engineered with full redundancy. This means that, in the event of a sub-circuit or full circuit failure, braking reverts to full hydraulic brake operation. ABSs are simple to troubleshoot. In most cases, when there is a poor braking complaint and no evidence of an ABS fault code, the problem source is outside of the ABS electronic circuit. Peak ABS modulator cycle rates can be as high as 20 times per second, so operation tends to be smoother and more pulse-free than pneumatic ABS. Aggressive panic stops will inevitably produce some noise which usually indicates the system is functioning normally. When brakes are noisy on application, the cause is usually outside of the ABS electronic circuit.

Task B1.13

Test, adjust, or replace antilock brake system (ABS) wheel speed sensors and tone/exciter rings/tooth.

Wheel speed sensors on hydraulic brake ABS, as with their pneumatic equivalents, are likely to produce most ABS complaints, due to their vulnerable location. Gapping complaints will usually be identified by the system self-diagnostics and are easy to remedy. Wheel speed sensors function using inductive pulse generation. A reluctor (known as a pulse generator, tone wheel, exciter ring or chopper wheel) cuts through a magnetic field inducing an AC current which acts as a signal to the ABS module. The module receives the AC electrical signal and correlates its frequency to an actual wheel speed value.

Wheel speed sensors should always be checked using the OEM recommended procedure. This can be by using a DMM in resistance mode or by a dynamic check using V-AC mode.

2. Mechanical System (6 Questions)

Task B2.1

Diagnose poor stopping, noise, pulling, grabbing, dragging, or pedal pulsation complaints caused by drum and disc brake mechanical assembly problems; determine needed repairs.

Hydraulic foundation brakes may use servo and nonservo principles. Servo action occurs when the action of one shoe is guided by the movement of the other, permitting both shoes to act as a single unit. Self-energization occurs when the shoes are driven into the drum and rotate fractionally as a pair with the drum, before grabbing.

Nonservo action is also a term generally used for certain drum/shoe-type brakes. Some original equipment manufacturers (OEMs) refer to these brakes as leading and trailing shoe brakes. Nonservo brakes have each shoe working independently of each other to stop the vehicle. They are separately anchored. When actuated by the wheel cylinder, the shoe pivots on the anchor and is forced against the drum.

Disc brakes are nonenergized, and they require more force to achieve the same braking effort as self-energized drum brakes. However, they have superior mechanical efficiency and are often used on the front axle brakes on vehicles. Front-wheel brakes perform a higher percentage of braking on most vehicles because of load transfer and suspension effects.

Grabbing and pulling can occur in the foundation brake assembly caused by broken hardware components, especially return springs, drum failures, malfunctions in the adjusting mechanism, and parking brake-related problems.

A primary cause of a pulsating pedal condition is warped disc brake rotors. Rotors warp due to overheating. Underspecifying the braking requirements of a vehicle, or machining rotors too thin at overhaul, can cause pedal pulsation and rotor warp. Thickness variations on disc brake rotors on different wheels can also cause pedal pulsation and loss of braking power.

Task B2.2

Inspect, resurface, reface, or replace brake drums or rotors.

Brake drums may be reused if they are within the manufacturer's specifications. The critical specifications are the maximum wear limit, machine limit, and maximum permissible diameter. Drums are measured with a drum gauge and should be checked for out of round, bell-mouthing, convexing, concaving, and taper. Drums should be inspected after measuring for heat checks and cracks and before they are machined.

Disc brake rotors may be reused if they are within the manufacturer's specifications. They should be measured for thickness with a micrometer and checked for parallelism and run-out with a dial indicator. If within machine limits, the rotor may be turned on a rotor lathe.

Task B2.3

Inspect, adjust, or replace drum brake shoes, linings, mounting hardware, adjuster mechanisms, and backing plates.

When a brake job is performed, the brake shoes, return springs, and fastening hardware are replaced. The friction face codes should be observed when replacing brake shoes. Most brake shoes today use bonded friction blocks, but riveted and bolted types are still in existence. Ensure that primary (leading) and secondary shoes are installed in their correct locations. Brake shoes are the first components to wear out in duo-servo brake systems.

Task B2.4

Inspect, or replace disc brake pads, and mounting hardware.

Most disc brake assemblies today have wear indicators that produce a squealing noise when the wear limit is exceeded. Servicing disc brake pads usually involves removing the caliper assembly, ensuring that float pins are not seized, backing off the automatic adjusting mechanism, and installing a new pair of brake pads. When a self-adjusting mechanism is used, care should be taken to ensure it is properly activated on reassembly. Whenever the brake pads are replaced, the brake rotors should be both measured and visually inspected.

Task B2.5

Inspect, adjust, and repair or replace in-wheel mechanical and hydraulic parking brake systems.

Cable actuated, rear parking brakes are used in some hydraulic brake school buses. The handbrake is connected by cable to a lever mechanism mounted within the brake foundation assembly. Cables used in this manner are vulnerable to seizure, especially when drivers do not regularly use them. Freeing steel cable in conduit can sometimes be accomplished by using penetration oil.

Task B2.6

Inspect, adjust, or replace driveline parking brake drums, rotors, bands, shoes, mounting hardware, and adjusters.

Cable-actuated, driveline parking brakes are used in some hydraulic brake systems especially in air-over-hydraulic brake systems. The unit is a band and rotor assembly mounted at the rear of the transmission; when engaged, the rear wheels are locked stationary by means of the driveshaft.

Task B2.7

Inspect, adjust, or replace driveline parking brake application system pedal, cables, linkage, levers, pivots, and springs.

A drive-line parking brake is incorporated as part of the drive shaft and is mechanically controlled by means of a cable actuated by the driver. The parking brake uses a lever mechanism mounted within the foundation assembly to force the brake shoes into the drums without any assist from the wheel cylinders.

Because the parking brake is actuated by a cable, it is vulnerable to seizure, especially when not used for extended periods. The parking brake cable may be freed up using penetrating oil, but it more often requires replacement. Driveline parking brake assemblies are usually mounted behind the transmission.

3. Power Assist Units and Miscellaneous (4 Questions)

Task B3.1

Diagnose poor stopping complaints caused by power brake booster (vacuum or hydraulic) problems; determine needed repairs.

In school bus hydraulic brake systems, the use of a vacuum or hydraulically assisted brake booster is required to reduce the foot effort that must be applied to the master cylinder to actuate the brakes. Vacuum assisted boosters are not found on today's medium duty on many current school buses using hydraulic brakes.

A hydraulic booster mechanism is powered by either the power steering pump or by a dedicated pump. The unit comprises an open center valve, reaction feedback mechanism, a large diameter boost or power piston, a reserve electric-powered pump, an integral flow control switch, and a power steering gear, operating in series. Vehicles with manual steering gear must use hydraulic boosters with a dedicated hydraulic pump.

Power brake boosters operate by assisting pedal effort with hydraulic pressure in proportion to pedal travel. Both single and dual diaphragm units are used. Malfunctioning hydro-boost units in school buses will usually require greatly increased brake pedal effort. The brake system cannot be effectively operated in this condition. Booster units may be repaired by overhaul or replacement.

Task B3.2

Inspect, test, repair, or replace power brake booster (vacuum or hydraulic), hoses and control valves.

In a hydromax power brake booster, if flow from the hydraulic pump is interrupted, an electric pump backs up the system. When replacing hydraulic hoses in the boost system, the lines must conform to the Society of Automotive Engineers (SAE) J189 standard.

Hydro-boost systems are bled by cranking the engine without starting it (the ignition system should be disabled) with reservoir filled with the recommended fluid. The refilling procedure may have to be repeated.

Task B3.3

Test, adjust, and replace brake stop light switch, bulbs, wiring, connectors, and warning devices.

Stoplights on hydraulic brake circuits may be actuated electromechanically or electrohydraulically. In either case, the switch simply grounds an electrical signal which closes the brake light circuit. The brake light circuit may be tested with a DMM.

C. Wheel Bearings Diagnosis and Repair (5 Questions)

Task C1   Remove and replace axle hub and wheel assembly.

The wheel assembly should be removed using a wheel dolly. The wheel seals should be removed using a heal bar or drift and a hammer. It is good practice to replace wheel seals each time they are removed from the hub. Unitized seals require replacement when removed because the rubber sealing surface is damaged on removal. Removing the wheel seal enables the bearing cone to be removed for cleaning and inspection; the cup may be inspected in the hub. If the bearing assembly has to be replaced, the cups may be driven out of their bores using an appropriately sized bearing driver. Alternatively, a mild steel drift and hammer may be used. The procedure is reversed to install cups. Never use a brass drift to install bearing cups.

Hubs should be prelubed when installing the wheel assembly. On nondriven axles, the bearing and hub assembly must be filled to a prescribed level in the calibrated inspection cover. In drive axles, the bearing and hub assembly is supplied with siphoned lubricant from the differential carrier.

Task C2   **Clean, inspect, lubricate, or replace wheel bearing assemblies; replace seals and wear rings.**

Wheel bearings should be inspected at each brake job and at each PM service that requires that the wheels be removed. Most current school buses use wet bearings; that is, they are lubricated with liquid lubricant, usually gear lube.

The use of taper roller bearings has become almost universal. A tapered wheel bearing assembly consists of a cone assembly and a cup or race. The cone assembly consists of tapered rollers mounted in a roller cage. The bearing race or cup is interference-fit to the hub. A pair of taper roller bearings support the load in each wheel hub. The cones and cups of taper roller bearings are not interchangeable; when damage is evident in either the cup or the cone, both must be replaced.

Bearings should be cleaned with solvent and air dried, ensuring that the cone is not spun out by the compressed air. Both the cone (rollers) and the cone should be inspected for spalling, galling, scoring, heat discoloration, and any sign of hard surface failure. When reinstalling wet bearings, they should be prelubed with the same oil to be used in the axle hub. When grease-packed bearings are used, the bearing cone must be packed with grease. This procedure may be performed by hand or by using a grease gun and cone packer. Grease-lubricated wheel bearings require the use of a high temperature axle grease. It is bad practice to pack wet bearings with axle grease and it may shorten bearing life by reducing lubrication efficiency.

Bearings fail primarily because of dirt contamination. This may be due to the conditions a vehicle is operated under or poor service practice. Lubrication failures caused by an inappropriate lubricant or lack of lubricant (caused by a failed wheel seal) also account for a large number of bearing failures: lack of lubricant will result in rapid failure and a bearing welded to an axle.

Many bearing failures are caused by brinelling. Brinelling appears as indentations across the bearing race. This condition occurs when the bearing is not rotating properly in the race. The condition is caused by impact loading or vibration.

Maladjusted bearings can fail rapidly, especially when the preload is high. The result of high preload on a bearing can cause the bearing to friction-weld to the axle. Overloaded bearings fail over time rather than rapidly.

Task C3   **Adjust axle wheel bearings in accordance with manufacturer's procedures.**

Most school buses use taper roller wheel bearings. Bearing adjustment usually requires that the bearing be seated to a specified torque value at the adjusting nut while rotating the wheel bearing; this part of the procedure is designed to seat the bearing with a preload. Next, the adjusting should be backed off to between one sixth and one third of a turn to locate the adjusting nut to the jam mechanism. Finally, the bearing end play must be measured with a dial indicator: the required specification is usually between 0.001 inch and 0.005 inch. End play must be present. If at least 0.001 inch end play is not present, the adjustment procedure should be repeated.

The bearing setting is locked in place on the axle spindle by a lock or jam nut that should be torqued to specification after the adjustment procedure. In some axles, a split forged or castellated nut and cotter pin are used to retain the wheel assembly. Bearing maladjustment can cause a wheel-off problem. The procedure for adjusting wheel ends is simple, but the consequences of maladjustment can be lethal. Always strictly observe the OEM recommended procedure.

Sample Test for Practice

Sample Test

Please note the letter and number in parentheses following each question. They match the overview in Section 4 that discusses the relevant subject matter. You may want to refer to the overview using this cross-referencing key to help with questions posing problems for you.

1. An air brake equipped school bus is in to have a service diaphragm on a spring brake assembly replaced. Technician A says the hold-off chamber must be "caged" before disconnecting any air line or hose. Technician B recommends that the service chamber be charged with air before removing the hold-off chamber. Who is right?
   A. A only
   B. B only
   C. Both A and B
   D. Neither A nor B   (A3.1 and A3.4)

2. Technician A says that one type of ABS ECU uses a series of light emitting diodes (LEDs) in a diagnostic window to indicate fault codes. Technician B says that ABS must provide redundancy back-up in the event of electronic failure. Who is right?
   A. A only
   B. B only
   C. Both A and B
   D. Neither A nor B   (A1.18)

3. All of the following are part of the foundation brake system **EXCEPT:**
   A. slack adjusters.
   B. pneumatic lines.
   C. brake shoes.
   D. brake chambers.   (A2.5)

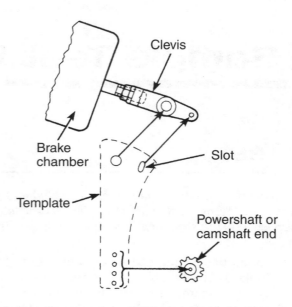

4. In the figure, the clevis is being reinstalled onto a pushrod. Technician A says that the clevis should be correctly positioned on the pushrod to achieve proper brake adjustment. Technician B says that the clevis on a brake chamber pushrod is keyed. Who is right?
   A. A only
   B. B only
   C. Both A and B
   D. Neither A nor B   (A2.1)

5. Technician A says that it is a good practice to change brake fluid whenever you perform a major brake repair on a bus with hydraulic brakes. Technician B says you should immediately replace contaminated brake fluid. Who is right?
   A. A only
   B. B only
   C. Both A and B
   D. Neither A nor B   (B1.10)

6. All of the following are part of an air compressor **EXCEPT:**
   A. piston rings.
   B. crankshaft.
   C. discharge valve.
   D. governor.   (A1.6)

7. A driver of a school bus equipped with air brakes complains of insufficient service brake application pressure when the brake pedal is depressed. Which of the following is the LEAST-Likely cause?
   A. Restricted air flow to the service brake chamber
   B. Improper adjustment of slack adjuster and chamber pushrod
   C. Grease on shoe linings/faces
   D. A ruptured diaphragm   (A1.1)

8. The first component(s) to wear out and need replacing on a typical duo-servo hydraulic braking system is the:
   A. wheel cylinder.
   B. drum shoes.
   C. anchor pins.
   D. return springs.   (B2.1)

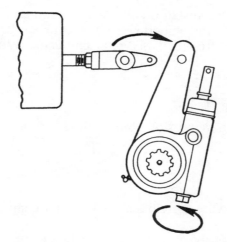

9. In the figure, a slack adjuster should be adjusted so that the brake chamber pushrod is:
   A. less than 90 degrees from the slack adjuster arm when fully applied.
   B. parallel to the slack adjuster arm when the brakes are fully applied.
   C. more than 90 degrees from the slack adjuster arm when fully applied.
   D. at a 90 degree angle from the slack adjuster arm when fully applied.   (A2.3)

10. What is the minimum number of air reservoirs installed on a FMVSS 121 compliant school bus?
    A. Three
    B. Six
    C. Nine
    D. Twelve   (A1.3)

11. Technician A says that all current school buses are required to have automatic slack adjusters. Technician B says that automatic slack adjusters increase the legal free stroke dimension. Who is correct?
    A. A only
    B. B only
    C. Both A and B
    D. Neither A nor B   (A2.1)

12. All of the following could cause a service brake to release sluggishly **EXCEPT:**
    A. misaligned actuation linkage.
    B. defective chamber return spring.
    C. weak brake shoe return springs.
    D. leaking service chamber.   (A2.1)

13. A technician has just replaced a hydraulic wheel cylinder. Technician A says that the pressure differential valve will require manual resetting. Technician B says that after manually resetting the pressure differential valve, you should replace the proportioning valve. Who is right?
    A. A only
    B. B only
    C. Both A and B
    D. Neither A nor B   (B1.7)

14. All of the following are parts of an automatic slack adjuster **EXCEPT:**
    A. worm gear.
    B. adjusting mechanism.
    C. actuator piston.
    D. clevis yoke.   (A2.1)

15. Technician A says you install the fail relay coil on an ABS system between the modulator assembly and a good ground. Technician B says power for the fail light flows through the normally closed relay contact(s). Who is right?
    A. A only
    B. B only
    C. Both A and B
    D. Neither A nor B   (A1.18)

16. All of the following are possible drum failure conditions **EXCEPT:**
    A. brinnelled.
    B. concave.
    C. scored.
    D. threaded.   (A2.8 and B2.2)

17. Which of the following describes a method of attaching a heavy-duty brake lining to the shoe?
    A. Integral
    B. Molded
    C. Staked
    D. Riveted   (A7.7)

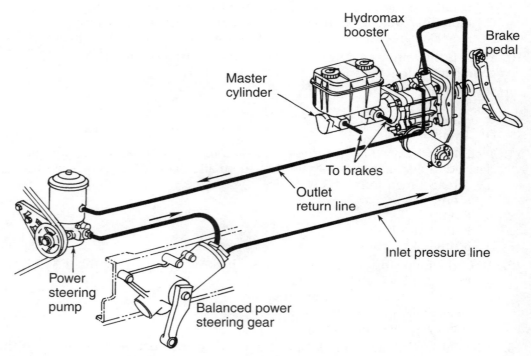

18. As shown in the figure, Technician A says that a hydrobooster and balanced steering gear can be integrated with the power steering pump circuit. Technician B says that a dedicated hydraulic power source can be used to power a hydrobooster. Who is right?
    A. A only
    B. B only
    C. Both A and B
    D. Neither A nor B   (B3.2)

19. A governor cut-out test is being performed. Technician A says cut-out should occur at 20–25 psi below governor cut-in. Technician B says to set governor cut-in before checking cut-out. Who is right?
    A. A only
    B. B only
    C. Both A and B
    D. Neither A nor B   (A1.6)

20. On a full-floating axle, what supports the vehicle's weight?
    A. The axle housing
    B. The axle shaft
    C. The outer axle bearing only
    D. The inner axle bearing only   (C1)

21. Technician A uses double flare tubing when replacing hydraulic brake lines. Technician B uses ISO tubing when replacing brake lines. Who is right?
    A. A only
    B. B only
    C. Both A and B
    D. Neither A nor B   (B1.5)

22. What is the LEAST-Likely cause of a blowing oil condition in an air compressor?
    A. A failed piston oil control ring
    B. Restrictions in the intake line or filter
    C. A defective head gasket
    D. Plugged sump drain ports   (A1.5)

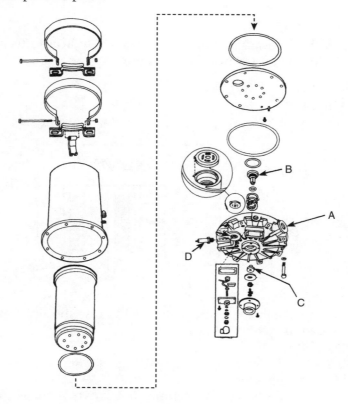

23. What is the component in the figure marked "C"?
    A. Unloader valve
    B. Purge valve
    C. Solenoid valve
    D. Desiccant pack   (A1.9)

24. What is the LEAST-Likely cause of poor service braking performance?
  A. Grease- or oil-contaminated shoes/linings
  B. Leaking quick-release valves
  C. Misadjusted slack adjuster/pushrod
  D. Leaking hold-off chamber   (A1.1)

25. A balanced air braking system is one in which:
  A. braking forces are proportional on all axles.
  B. braking pressure reaches each actuator simultaneously.
  C. each axle receives braking force in a sequential manner, starting with that
      closest to the governor.
  D. each wheel receives graduated braking force over a wide range of air pressure.
                                                                            (A1.1)

26. All of the following could cause excessive leakage in an air brake system when the
    service brakes are applied **EXCEPT:**
  A. a leaking service chamber diaphragm.
  B. a leaking hose.
  C. a defective governor.
  D. a defective relay valve.   (A1.2)

27. Pushrod free stroke is being measured. Technician A says that the applied stroke
    should use an 80-psi brake application. Technician B says the brake clevis and
    pushrod are adjusted to achieve the proper free stroke. Who is right?
  A. A only
  B. B only
  C. Both A and B
  D. Neither A nor B   (A2.3)

28. Technician A checks bearings end play with a dial indicator. Technician B sets
    wheel-end with a slight preload. Who is right?
  A. A only
  B. B only
  C. Both A and B
  D. Neither A nor B   (C3)

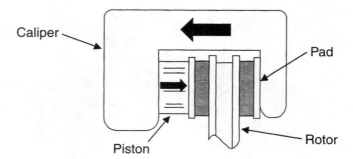

29. On the air disc brake in the figure, what could cause short piston pad life?
  A. Piston seized in its bore
  B. Damaged rotor
  C. Defective brake hose
  D. Seized caliper slide pins   (A2.7)

30. You are servicing an ABS system. Technician A says that wheel speed sensors
    produce an AC electrical signal. Technician B says that when an ABS system is
    disabled, the vehicle must legally revert to standard braking mode. Who is right?
  A. A only
  B. B only
  C. Both A and B
  D. Neither A nor B   (A1.19 and A1.20)

31. A slack adjuster is being adjusted. Technician A says wear between the clevis pin and yoke will cause the stroke to be too long. Technician B says that an automatic slack adjuster never should be adjusted. Who is right?
    A. A only
    B. B only
    C. Both A and B
    D. Neither A nor B   (A2.3)

32. Technician A says that hygroscopic brake fluid is required for use in all school bus brake systems. Technician B says that hygroscopic brake fluids easily absorb moisture from air and should not be exposed to air. Who is right?
    A. A only
    B. B only
    C. Both A and B
    D. Neither A nor B   (B1.10)

33. What should you not do, when replacing the diaphragm on a service brake chamber?
    A. Remove the hold-off clamps.
    B. Cage the spring brake.
    C. Loosen the service chamber clamps by tapping them with a mallet.
    D. Exhaust the air pressure.   (A2.2)

34. An aftercooler air dryer:
    A. requires periodic changing of its oil filter.
    B. condenses hot air to remove moisture.
    C. receives cool air from the compressor.
    D. should be drained daily.   (A1.9)

35. Where is a quick-release ratio valve mounted?
    A. Close to the front service chambers
    B. Close to the bus protection valve
    C. In the middle of a tandem axle trunnion
    D. On the outside backwall of the tractor next to the gladhands   (A1.13)

36. Technician A says it's okay to machine oversize drums as long as the drum manufacturer's recommendation for machining dimensions are followed. Technician B says that it is good practice to turn new drums because they distort in storage. Who is right?
    A. A only
    B. B only
    C. Both A and B
    D. Neither A nor B   (B2.2)

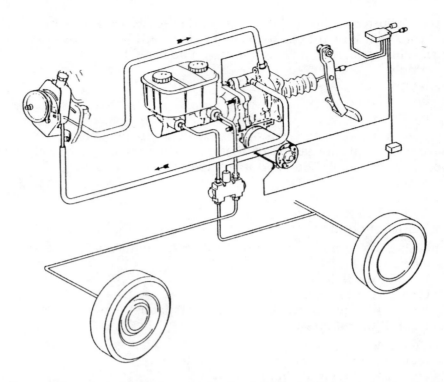

37. In the figure, what force is used to apply the foundation brake wheel cylinders?
    A. Hydraulic, from the power steering reservoir
    B. Mechanical
    C. Electrical
    D. Hydraulic, from the master cylinder   (B3.2)

38. What are school bus air foundation brake components mounted to?
    A. Backing plates
    B. A spider
    C. Axle flange
    D. Anchor pins   (A2.5)

39. A wheel speed sensor fails in a four-channel school bus ABS air system. Which of
    the following is the Most-Likely outcome?
    A. The vehicle should be pulled over immediately and repaired.
    B. Foundation brakes function at 75 percent potential.
    C. The ABS ECU shuts down ABS management of braking.
    D. The failed portion of the ABS circuit defaults to standard braking.   (A2.5)

40. Technician A says that the critical length of a slack adjuster is the dimension
    between the center of the S-camshaft to the center of the clevis pin. Technician B
    says that when replacing a slack adjuster, increasing the critical length is
    acceptable because it increases brake torque at the wheel. Who is right?
    A. A only
    B. B only
    C. Both A and B
    D. Neither A nor B   (A2.3)

41. The retraction spring breaks in the foundation assembly of a school bus S-cam
    system. What is the Most-Likely consequence?
    A. Delayed application
    B. Reduced brake torque
    C. Parking brake lag
    D. Dragging brakes   (A2.1)

42. The S-cam on one foundation brake fitted rolls over. Which of the following is LEAST-Likely to be the cause?
    A. Brakes out of adjustment
    B. Worn S-cam bushings
    C. Broken anchor spring
    D. Grease on S-cam (A2.4)

43. You are servicing a piggyback spring brake assembly. Technician A says no attempt should be made to repair any part of the piggyback assembly, but that it should be completely replaced as a unit. Technician B says that the spring chamber should be disarmed, before discarding. Who is right?
    A. A only
    B. B only
    C. Both A and B
    D. Neither A nor B (A2.2)

44. How is the relay section of a dual circuit, air brake foot valve actuated if the valve is functioning properly?
    A. Air pressure from the secondary reservoir
    B. Air pressure from the primary section of the foot valve
    C. Mechanically, by direct foot pressure
    D. Mechanically, by the primary piston (A1.10)

45. A school bus equipped with air brakes completely loses the primary circuit air reservoir. Under these conditions, how is the relay section of a dual circuit, air brake foot valve actuated?
    A. Air pressure from the secondary reservoir
    B. Air pressure from the primary section of the foot valve
    C. Mechanically, by direct foot pressure
    D. Mechanically, by the primary piston (A1.10)

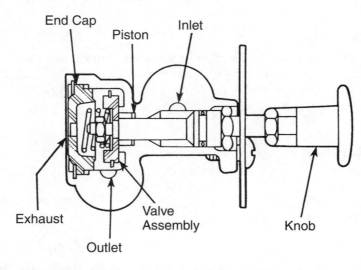

46. In the figure, when the park control valve is pulled out:
    A. the parking brake released.
    B. air builds up in the piggyback.
    C. it will not reopen until the treadle valve is depressed.
    D. air in the hold-off chambers is exhausted. (A1.14)

47. Technician A says when disk brakes are not properly adjusted, this reduces brake mechanical advantage. Technician B says that a properly adjusted S-cam brake is one where the angle between the slack adjuster and the chamber pushrod is 90 degrees when the brakes are released. Who is right?
   A. A only
   B. B only
   C. Both A and B
   D. Neither A nor B   (A2.3 and A2.6)

48. A school bus with drum brakes is being serviced. Technician A says that the brakes may be disk-actuated. Technician B says that the brakes may be cam-actuated. Who is right?
   A. A only
   B. B only
   C. Both A and B
   D. Neither A nor B   (A2.1)

49. Technician A says that when a front disc-rear drum hydraulic system is used, an anticompounding valve is also used. Technician B says that a proportioning valve is used in a front disc-rear drum non-ABS braking system. Who is right?
   A. A only
   B. B only
   C. Both A and B
   D. Neither A nor B   (B1.6)

50. A driver complains of dragging brakes in an air brake equipped school bus. Technician A says a leaking hold-off diaphragm could cause the problem. Technician B says a broken return spring in the service chamber could be the cause. Who is right?
   A. A only
   B. B only
   C. Both A and B
   D. Neither A nor B   (A1.1)

51. While road testing a school bus, after a master cylinder overhaul, the brakes drag. Technician A says that the compensation ports may be blocked. Technician B says that excessive pedal free play could be the cause. Who is right?
   A. A only
   B. B only
   C. Both A and B
   D. Neither A nor B   (B1.4)

52. If a technician fails to install master cylinder reservoir cover diaphragm, which of these will it cause?
   A. Fluid contamination
   B. Loss of pressure
   C. Seal erosion
   D. Air in the fluid   (B1.4)

53. You are inspecting a four-wheel disc hydraulic brake system. Technician A says sticking caliper pistons may cause brake drag. Technician B says that rotors machined below specified minimum thickness will rapidly warp due to overheating. Who is right?
   A. A only
   B. B only
   C. Both A and B
   D. Neither A nor B   (B1.9)

54. The diameter on a driveline parking brake drum is greater at the edges of the drum than in the center. Which of the following describes this condition?
    A. Bell-mouth drum
    B. Concave brake drum
    C. Drum out of round
    D. Convex drum
    (B2.6)

55. When a school bus with hydraulic disc brakes makes panic-stop brake applications, brake fade occurs. Technician A says this could be caused by low fluid level. Technician B says brake drums worn or machined beyond the maximum diameter could cause this condition. Who is right?
    A. A only
    B. B only
    C. Both A and B
    D. Neither A nor B
    (B2.1)

56. You are inspecting brake pads on a school bus equipped with hydraulic disc brakes. Technician A says you must remove the wheel in most cases to measure pad thickness. Technician B says if the brakes were applied shortly before measuring the lining to rotor clearance, this clearance will be greater than specified. Who is right?
    A. A only
    B. B only
    C. Both A and B
    D. Neither A nor B
    (B2.4)

57. A school bus with a hydrovac brake booster requires excessive brake pedal effort. Technician A says the one-way check valve may be restricted in the vacuum hose to the hydrovac unit. Technician B says a sticking power piston in the hydrovac unit could be the cause. Who is right?
    A. A only
    B. B only
    C. Both A and B
    D. Neither A nor B
    (B3.1)

58. Cable-actuated parking brakes on hydraulic brake systems are found in:
    A. most diagonally split brake systems.
    B. all disc brake systems.
    C. all hydraulically operated systems.
    D. some air-over-hydraulic parking systems.
    (B2.5)

59. What connects a slack adjuster to an S-camshaft?
    A. Clevis pin
    B. Splines
    C. Cotter pin
    D. Bolts
    (A2.3)

60. Which of the following dimensions would be correct when measuring applied stroke on an automatic slack adjuster?
    A. 0.25"
    B. 0.75"
    C. 1.25"
    D. 2.00"
    (B2.3)

6 Additional Test Questions for Practice

Additional Test Questions

Please note the letter and number in parentheses following each question. They match the overview in Section 4 that discusses the relevant subject matter. You may want to refer to the overview using this cross-referencing key to help with questions posing problems for you.

1. Which of the following is a function of a two-way check valve in an air brake system?
   A. It ensures that air is applied in only one direction at one time.
   B. If one circuit loses pressure, the two-way check valve permits the functioning circuit to supply air to both circuits.
   C. It alerts the driver as to which direction the air in a particular line is traveling at a given time.
   D. It ensures proper levels for governor cut-off pressures. (A1.11)

2. The majority of air brake applications are made at an application pressure of:
   A. 100 psi or less.
   B. 80 psi or less.
   C. 40 psi or less.
   D. 20 psi or less. (A1.1)

3. Technician A says that the park and emergency braking system is a separate air circuit, isolated from the service air system. Technician B says that hold-off chambers produce pneumatic force in the opposite direction to the service brake chambers. Who is right?
   A. A only
   B. B only
   C. Both A and B
   D. Neither A nor B (A3.2)

4. You are servicing wheel bearings on a school bus. Technician A says that you use a high quality high, temperature grease when repacking is necessary. Technician B says it's okay to switch wheel bearings from one wheel to another. Who is right?
   A. A only
   B. B only
   C. Both A and B
   D. Neither A nor B (C2)

5. Which of the following is known as brake compounding?
   A. Simultaneous application of air to front and rear brakes
   B. Driving the vehicle with the parking brake applied
   C. Releasing the air in the spring brake chamber to allow the brakes to be applied
   D. Simultaneous application of service and spring brake forces to the foundation brakes (A1.11)

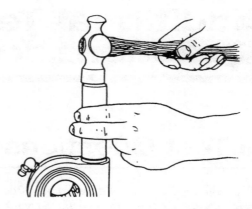

6. As shown in the figure, you are installing a seal on a worm gear of a slack adjuster. Technician A says to install the lip of the seal facing toward the outside of the adjuster. Technician B says not to drive the seal after it contacts the bottom of the bore. Who is right?
   A. A only
   B. B only
   C. Both A and B
   D. Neither A nor B   (A2.3)

7. Technician A says that power brake boosters are used on hydraulic brake systems. Technician B says that power brake boosters are used on air-over-hydraulic systems. Who is right?
   A. A only
   B. B only
   C. Both A and B
   D. Neither A nor B   (B3.2)

8. Technician A says that cam-actuated foundation brakes are the most common air brake system on current school buses. Technician B says that slack adjusters and S-cams convert linear motion into brake torque. Who is right?
   A. A only
   B. B only
   C. Both A and B
   D. Neither A nor B   (A2.4)

9. The discharge valve in an air compressor:
   A. directs air into the primary tank.
   B. prevents compressed air in the discharge line from returning to the compressor cylinder.
   C. prevents the pressure in the compressor cylinder from becoming too high.
   D. allows compressed air to pass into the governor.   (A1.6)

10. The pressure differential valve in a hydraulic brake system can also be known as the:
   A. pressure warning valve.
   B. proportioning valve.
   C. metering valve.
   D. system effectiveness indicator regulating valve.   (B1.7)

11. Technician A says that the pulse cycle rate in a school bus air ABS system can be as fast as 7 times per second. Technician B says that hydraulic ABS systems pulse cycle at up to 20 times per second. Who is right?
   A. A only
   B. B only
   C. Both A and B
   D. Neither A nor B   (A1.18 and B1.11)

12. Which of the following is NOT a method of attaching liner friction facings to brake shoes?
    A. Adhesive bonding
    B. Screws and nuts
    C. Rivets
    D. Welding   (A2.7)

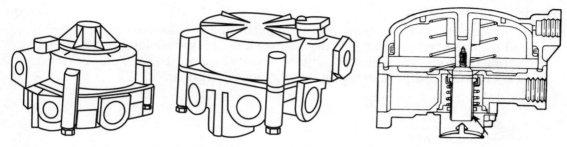

13. The above figure shows different types of:
    A. quick release valve.
    B. ratio valve.
    C. tractor protection valve.
    D. relay valve.   (A1.13)

14. The driver of an air brake equipped school bus traveling down a highway makes a full service application of the brakes. At the time of the service application, which brake chambers on the vehicle are charged with air?
    A. Service chambers only
    B. Hold-off chambers only
    C. Service and hold-off chambers
    D. All chambers are evacuated.   (A3.1)

15. In the S-cam foundation brake, which of the following components is responsible for multiplying brake torque?
    A. The chamber pushrod
    B. The slack adjuster
    C. The camshaft
    D. The chamber mounting plate   (A2.3)

16. Technician A says that it is friction on the drum/rotor that retards vehicle motion. Technician B says braking means converting the energy of movement into heat energy. Who is right?
    A. A only
    B. B only
    C. Both A and B
    D. Neither A nor B   (A2.8)

17. Technician A says that one of the factors affecting air timing is the size of the air lines. Technician B says that if one service chamber has an air fitting that has a lower flow area, then the air timing will be affected. Who is right?
    A. A only
    B. B only
    C. Both A and B
    D. Neither A nor B   (A1.1)

18. A school bus with a hydraulic braking system is being overhauled. Technician A uses a vacuum gauge to test master cylinder output. Technician B says that all the wheel cylinders must be replaced when an overhaul is being performed. Who is right?
    A. A only
    B. B only
    C. Both A and B
    D. Neither A nor B   (B1.4)

19. Technician A says that a slack adjuster must be adjusted so the angle between it and the chamber pushrod is 90 degrees when the brakes are fully applied. Technician B says that the slack adjuster should be adjusted with the shortest possible stroke, without dragging the brakes. Who is right?
    A. A only
    B. B only
    C. Both A and B
    D. Neither A nor B   (A2.3)

20. Which of these is the LEAST-Likely cause of wheel bearing failure?
    A. Overloading
    B. Contamination
    C. A damaged race
    D. Improper lubricant   (C2)

21. How many teeth are on a typical wheel speed sensor reluctor used in a school bus application?
    A. 16
    B. 32
    C. 100
    D. 256   (A1.20)

22. What should be used when flushing a hydraulic brake system?
    A. Water
    B. Mineral spirits
    C. Gasoline
    D. Alcohol   (B1.10)

23. System pressure in a school bus air brake system builds to 150 psi and holds at that pressure with the compressor not going into unloading cycle. Technician A says that the safety pop-off valve is defective and should be replaced. Technician B says that the first step in troubleshooting the condition should be to check the governor unloader line. Who is right?
    A. A only
    B. B only
    C. Both A and B
    D. Neither A nor B   (A1.6 and A1.8)

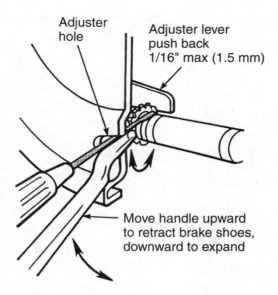

Adjuster hole

Adjuster lever push back 1/16" max (1.5 mm)

Move handle upward to retract brake shoes, downward to expand

24. What is being done to the hydraulic brakes in the figure?
    A. Removing the shoes
    B. Removing the drum
    C. Brake adjustment
    D. Setting preload   (B2.3)

25. A hydromax power brake booster is being serviced. Technician A says that if normal flow from the hydraulic brake pump is interrupted, the electric pump provides power for reserve stops. Technician B says unassisted braking is possible if the power and reserve systems fail, but the braking will be extremely difficult. Who is right?
    A. A only
    B. B only
    C. Both A and B
    D. Neither A nor B   (B3.1)

26. A school bus ABS system is described as 4S/4M. Which of the following has to be true?
    A. Two of the wheels are modulated in pairs.
    B. The vehicle has 6 tires.
    C. The front brakes are not modulated.
    D. All four wheels are modulated.   (A1.20)

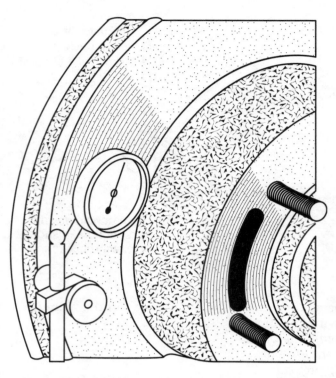

27. An air disc brake system is being inspected. Technician A says that rotor thickness is being checked in the figure. Technician B says rotor run out is being measured. Who is right?
    A. A only
    B. B only
    C. Both A and B
    D. Neither A nor B   (A2.8)

28. When a gear driven air compressor goes into unloaded cycle, which of the following would be true?
    A. The discharge valves are held open by the unloader mechanism.
    B. The inlet valves are held open by the unloader mechanism.
    C. The governor exhausts the unloader signal.
    D. The governor cut-in pressure is achieved.   (A1.6)

29. The machine limit for a brake drum is specified at 0.090 inches over and the maximum legal wear limit 0.120 inches over. Technician A says that a drum that measures 0.105 inches over during a PM inspection should be discarded. Technician B says that the drum may be turned but care should be taken not to exceed 0.120 inches over. Who is right?
    A. A only
    B. B only
    C. Both A and B
    D. Neither A nor B   (A2.8)

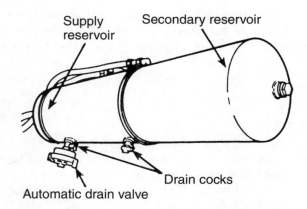

30. In the figure, what must there be between the supply and secondary reservoirs?
    A. Internal baffle and one-way check
    B. Internal baffle and two-way check
    C. A safety pop-off
    D. A spitter valve   (A1.3)

31. During a wheel seal repair, the brake friction facings are observed to be saturated with oil and glazed to a mirror finish. Technician A says that the linings should be cleaned with liquid detergent and power washing. Technician B says that the brake linings should be replaced, along with those on the other side of the axle.
    Who is right?
    A. A only
    B. B only
    C. Both A and B
    D. Neither A nor B   (A2.7)

32. Never use brake fluid out of a container:
    A. that is made out of glass.
    B. that has been air sealed.
    C. that has been used to store any other liquid.
    D. that has had brake fluid in it for more than six months.   (B1.10)

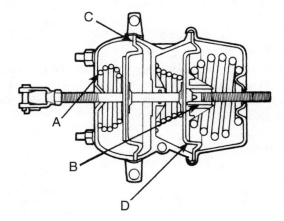

33. What is the part marked "A" in the figure?
    A. Hold-off diaphragm
    B. Power spring
    C. Service retraction spring
    D. Pressure plate   (A2.2)

34. A wheel speed sensor from an ABS system is being inspected. Technician A says that as the toothed wheel rotates, the sensor generates a DC signal. Technician B says that the ABS-ECU reads the frequency generated by the wheel speed sensor. Who is right?
    A. A only
    B. B only
    C. Both A and B
    D. Neither A nor B   (A1.20)

35. After installing an air compressor, it is discovered that the steel tube that inserts into the front of the compressor drive shaft has not been installed. Technician A says not to worry because this is a guide pin, used only to align the compressor during installation. Technician B says that if the compressor is not removed and the tube installed, it will fail due to lack of lubricant. Who is right?
    A. A only
    B. B only
    C. Both A and B
    D. Neither A nor B   (A1.4)

36. Technician A says that a hydraulic disc brake caliper can have two pistons. Technician B says that a disc brake caliper can have four pistons. Who is right?
    A. A only
    B. B only
    C. Both A and B
    D. Neither A nor B   (B2.6)

37. A return line is being installed to a hydraulic power booster. Technician A says the line must be hydraulic hose conforming to SAE J189. Technician B uses a pressure bleeder to bleed the hydro-boost circuit. Who is right?
    A. A only
    B. B only
    C. Both A and B
    D. Neither A nor B   (B3.2)

38. Technician A says both front and rear wheel bearings are adjustable. Technician B says you use a brass drift to remove a rear wheel bearing race. Who is right?
    A. A only
    B. B only
    C. Both A and B
    D. Neither A nor B   (C2)

39. Which of the following best describes the action of hydraulic single piston disc brakes?
    A. Servo action
    B. Nonservo action
    C. Energizing action
    D. Action and reaction   (B1.9)

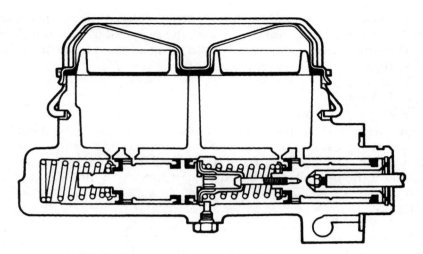

Cross-section of a typical dual master cylinder

40. As shown in the figure, after a brake application, what component returns the master cylinder pistons to the unapplied position?
    A. Secondary piston
    B. Compensating port
    C. Sliding rubber seals
    D. Return springs   (B1.4)

41. When discussing an S-cam-actuated air brake system, Technician A says the service chamber retraction spring in combination with the brake shoe return spring releases the brakes after a service application. Technician B says the clevis yoke threads onto the slack adjuster arm. Who is right?
    A. A only
    B. B only
    C. Both A and B
    D. Neither A nor B   (A2.2)

42. Technician A says that on brakes with vented rotors, the inboard pad is sometimes thicker than the outboard pad. Technician B says that on brakes with solid rotors, the inboard and outboard pads are usually the same thickness. Who is right?
    A. A only
    B. B only
    C. Both A and B
    D. Neither A nor B   (B2.4)

43. A school bus has a problem with system air pressure rising above normal. Technician A says the check valve on the governor could be stuck closed. Technician B says there could be too much clearance at the compressor unloading valves. Who is right?
    A. A only
    B. B only
    C. Both A and B
    D. Neither A nor B   (A1.6)

44. What force in spring brake chambers is being used to brake the vehicle during service braking?
    A. Spring brake power spring
    B. Service air
    C. Service chamber retraction spring
    D. Hold-off air   (A2.2)

45. A wheel bearing with a unitized seal is being replaced. Technician A says any time a wheel with a unitized seal is removed, the seal *must* be replaced. Technician B says that wheel seals should be removed using a pair of heal bars. Who is right?
   A. A only
   B. B only
   C. Both A and B
   D. Neither A nor B   (C2)

46. Technician A says that the air brake system low-pressure light should illuminate when system pressure drops below 60 psi. Technician B says that once system pressure drops below 60 psi, there is insufficient pressure in the hold-off chambers to hold off the spring brakes and they start to apply. Who is right?
   A. A only
   B. B only
   C. Both A and B
   D. Neither A nor B   (A1.12)

47. Technician A says that when performing a brake overhaul on a hydraulic rear drum brake system, you replace the return springs. Technician B says the maximum allowable out-of-round specification on a brake drum should be 0.025 inch. Who is right?
   A. A only
   B. B only
   C. Both A and B
   D. Neither A nor B   (B2.3)

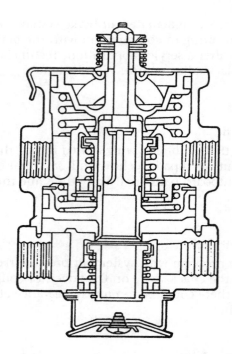

48. The dual circuit application valve in the figure is being discussed. Technician A says that the primary piston is actuated mechanically in both normal and failure modes. Technician B says that in the event of a complete primary circuit failure, the relay piston is mechanically actuated by downward movement of the primary piston. Who is right?
   A. A only
   B. B only
   C. Both A and B
   D. Neither A nor B   (A1.10)

49. Technician A says that compressor discharge air temperature is often 300° F. Technician B says that most air compressors are liquid cooled. Who is right?
    A. A only
    B. B only
    C. Both A and B
    D. Neither A nor B   (A1.5)

50. A full-floating wheel assembly is being inspected. Technician A says that on disc wheels, the wheel and hub are combined into a single unit. Technician B says a wheel bearing nut may be locked in place by either a lock nut, split forged nut, or castellated nut and cotter pin. Who is right?
    A. A only
    B. B only
    C. Both A and B
    D. Neither A nor B   (C1)

51. During air brake diagnosis the application valve is suspected to be the cause of slow apply. When diagnosing the brake application valve, Technician A says the air pressure at the delivery port should be proportional to treadle mechanical travel. Technician B says with the brakes applied, a 1 inch (2.54 cm) bubble in 3 seconds at the brake application valve exhaust port indicates excessive leakage. Who is right?
    A. A only
    B. B only
    C. Both A and B
    D. Neither A nor B   (A1.10)

52. Technician A says master cylinder components may be washed in petroleum-based solvent. Technician B says an aluminum master cylinder bore may be honed if it is scored. Who is correct?
    A. A only
    B. B only
    C. Both A and B
    D. Neither A nor B   (B1.4)

53. When discussing quick-release valve diagnosis, Technician A says that with the service brakes applied, acceptable leakage at the quick-release valve exhaust port is a 3 inch (2.54 cm) bubble in 1 second. Technician B says that when the service brakes are released, the air pressure in the front brake chambers should be exhausted immediately. Who is right?
    A. A only
    B. B only
    C. Both A and B
    D. Neither A nor B   (A1.1 and A1.13)

54. During a brake inspection, a technician tests the air supply system and finds that buildup is slow. Which of these is the Most-Likely cause?
    A. Worn compressor rings
    B. Leak in a service brake chamber
    C. Air leak in the rear suspension
    D. Restriction in the governor line   (A1.2)

55. An automatic reservoir drain valve does which of the following?
    A. It opens when the reservoir pressure reaches the governor pressure.
    B. It is mounted in a threaded fitting near the top of the reservoir.
    C. It opens by decreasing supply reservoir pressure.
    D. It drains water if sump pressure is 2 psi higher than reservoir pressure.   (A1.3)

56. When you are adjusting the tension of the compressor belt, and you notice contact at the bottom of the pulley groove with the belt, you should:
    A. decrease tension to original specification level.
    B. measure the deflection pulleys.
    C. replace the belt.
    D. ignore it because some applications require additional contact area.   (A1.4)

57. You are replacing a single-cylinder air compressor gear driven by the engine. Technician A says it may be necessary to time the compressor to the engine. Technician B says two-cylinder compressors do not usually require timing to the engine. Who is right?
    A.. A only
    B. B only
    C. Both A and B
    D. Neither A nor B   (A1.5)

58. The dual wheel assembly of a school bus drive axle is being reinstalled. Technician A says to put a light coating of lubricant on the spindle and place a seal protector over the threads. Technician B says rear drive wheel hub assemblies are lubricated by oil from the differential carrier. Who is right?
    A. A only
    B. B only
    C. Both A and B
    D. Neither A nor B   (C1)

59. When testing and adjusting the governor cut-out and cut-in pressures, which of the following statements is true?
    A. Cut-out pressure is increased by rotating the adjusting screw counterclockwise.
    B. Turning the adjusting screw clockwise increases cut-in pressure.
    C. Turn the adjustment screw 1/4 turn to change cut-in or cut-out pressure 7 psi.
    D. Restricting the air line from the supply reservoir to the governor decreases cut-out pressure.   (A1.6)

60. Technician A says that daily draining of air reservoirs not equipped with automatic drain valves is recommended. Technician B says that automatic drain valves and moisture removing devices should be checked periodically for proper operation. Who is right?
    A. A only
    B. B only
    C. Both A and B
    D. Neither A nor B   (A1.3)

61. What is the LEAST-Likely cause of water in air reservoirs?
    A. Dented or kinked air lines
    B. Lack of periodic drainage of supply tanks
    C. Oil saturated air dryer dessicant
    D. Continuous operation in humid conditions   (A1.9)

62. A school bus with air brakes has its parking brakes applied and one wheel jacked up so it is clear of the shop floor. The wheel rotates easily when turned by hand. Which of the following would be the Most-Likely cause?
    A. Worn brake linings
    B. Broken power spring
    C. Leaking hold-off diaphragm
    D. Brake out of adjustment   (A2.2)

63. When diagnosing service brake relay valve problems, all of the following apply **EXCEPT:**
    A. inlet valve leakage is tested with the service brakes released.
    B. signal air pressure should equal inlet pressure.
    C. exhaust valve leakage is tested with the service brakes applied.
    D. service application air is discharged through the exhaust port.   (A1.15)

64. When considering a school bus service brake relay valve:
    A. air lines connect the delivery ports to the service brake chambers.
    B. when the brakes are released, the inlet valve is open.
    C. when the brakes are released, the exhaust valve is closed.
    D. the relay valve is in a balanced position when pressure in the brake chambers equals reservoir pressure.   (A1.13)

65. The lock collar is seized downward on a manual slack adjuster and will not retract to lock the adjustment. Technician A says that the collar should be staked with a chisel to hold the adjustment. Technician B says that the slack adjuster must be replaced. Who is right?
    A. A only
    B. B only
    C. Both A and B
    D. Neither A nor B   (A2.3)

66. Technician A says that air governor cut-in pressure should be adjusted before the cut-out pressure. Technician B says that the difference between governor cut-out and governor cut-in should be between 20 and 25 psi. Who is right?
    A. A only
    B. B only
    C. Both A and B
    D. Neither A nor B   (A1.6)

67. Governor cut-out is set at 125 psi on a school bus air system. However, governor cut-in occurs at 115 psi. Which of the following would be Most-Likely to repair the problem?
    A. Adjust governor cut-in to 105 psi.
    B. Check the operation of the pop-off valve.
    C. Replace the compressor.
    D. Replace the governor.   (A1.6)

68. While discussing single and double check valves, Technician A says a single check valve may be located between the primary and secondary reservoirs. Technician B says a double check valve routes air pressure from the lowest of two pressure sources to a single outlet. Who is correct?
    A. A only
    B. B only
    C. Both A and B
    D. Neither A nor B   (A1.8)

69. The low-pressure warning light on a vehicle dash does not go out after startup. Technician A says that the problem may be caused by a loss of one section of the dual system. Technician B says that the problem may be caused by a short circuit. Who is right?
    A. A only
    B. B only
    C. Both A and B
    D. Neither A nor B   (A1.15)

70. Technician A says that an inoperative warning light below 60 psi could be caused by a defective warning light bulb. Technician B says that an inaccurate dash pressure gauge can cause a warning light to fail. Who is right?
    A. A only
    B. B only
    C. Both A and B
    D. Neither A nor B   (A1.15)

71. All of the following can be causes of brake torque imbalance **EXCEPT:**
    A. oil-soaked linings.
    B. nonuniform adjustment.
    C. nonuniform chamber sizes.
    D. an overloaded vehicle.   (A2.1)

72. The brake linings are worn and there is excessive clearance between the linings and the brake drum on a S-cam-type foundation brake. Technician A says this condition may cause the cam to roll over during a brake application. Technician B says this condition may cause loss of steering control under braking. Who is right?
    A. A only
    B. B only
    C. Both A and B
    D. Neither A nor B   (A2.1)

73. When assembling an S-camshaft all of the following apply **EXCEPT:**
    A. both camshaft seal lips must face toward the slack adjuster.
    B. use the proper driver when installing camshaft seals.
    C. lubricate the cam profile with petroleum jelly.
    D. measure camshaft radial movement.   (A2.4)

74. All of the following are parts of an air disc brake assembly **EXCEPT:**
    A. cam actuator.
    B. torque plates.
    C. calipers.
    D. rotors.   (A2.6)

75. New brake linings are being installed on a school bus. Technician A says you use the same fastener sequence with bolted linings as with riveted ones. Technician B always greases the actuator roller profiles before installing new shoes/linings. Who is right?
    A. A only
    B. B only
    C. Both A and B
    D. Neither A nor B   (A2.7)

76. Technician A says that severe drum heat checking will cause rapid lining wear. Technician B says that brake drums may be machined 0.130 inch over nominal diameter. Who is right?
    A. A only
    B. B only
    C. Both A and B
    D. Neither A nor B   (A2.8)

77. What can extreme rotor run out produce in a hydraulic brake system?
    A. Low brake output
    B. Pedal pulsation
    C. Loss of brake fluid
    D. Loss of directional control   (B2.2)

78. Which of the following is the LEAST-Likely cause of a parking brake that will not hold?
    A. Slack adjuster free stroke adjustment
    B. Slack adjuster alignment
    C. Worn foundation brake linings
    D. Weak spring brake chamber springs   (A2.1)

79. Which of the following is regarded as bad practice when servicing a typical spring brake assembly?
    A. Caging the power spring
    B. Replacing the service diaphragm
    C. Replacing the hold-off diaphragm
    D. Loosening the service chamber clamps to align the air ports   (A2.2)

80. All of the following could cause a hydraulic ABS system fault **EXCEPT:**
    A. shorted solenoid coils.
    B. leaking hydraulic lines.
    C. blocked exhaust solenoid ports.
    D. an open circuit in the ECU.   (B1.12)

81. Technician A says that ATC functions by applying brake pressure to a drive axle wheel when its velocity exceeds the mean (average) velocity of the other drive wheel. Technician B says that full authority ATC functions on all four wheels of a school bus. Who is right?
    A. A only
    B. B only
    C. Both A and B
    D. Neither A nor B   (A1.18)

82. Which of the following is the most commonly used ABS system on school buses?
    A. Eight-channel system
    B. Six-channel system
    C. Four-channel system
    D. Two-channel system   (A1.18)

83. Which of the following conditions can cause brake pedal fade in a hydraulic brake system?
    A. A seized brake caliper piston
    B. Brake drum machined beyond its limit
    C. Leakage past the master cylinder cups
    D. Air in the hydraulic system

84. What should the inside of a new hydraulic brake line be c
    A. Soap and water
    B. Isopropyl alcohol
    C. Mineral spirits
    D. Gasoline

85. On a school bus equipped with disc/drum brakes, the pedal pressure is applied. Which of the following wo
    A. Proportioning valve
    B. Pressure differential valve
    C. Metering valve
    D. Residual check valve

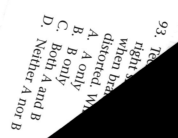

86. A hydraulic brake system with a metering valve is being bled. Technician A applies a device to hold the metering valve open when he uses a pressure bleeder to bleed the system. Technician B bleeds the brakes manually and does not hold the metering valve open. Who is right?
    A. A only
    B. B only
    C. Both A and B
    D. Neither A nor B   (B1.10)

87. The diameter on a driveline parking brake drum is greater in the center than at the edges. Which of the following describes this condition?
    A. Bell-mouth drum
    B. Concave drum
    C. Drum out of round
    D. Convex drum   (B2.6)

88. What action would occur if the reserve electric motor for the hydraulic power brake system failed?
    A. No reserve stops without the engine running.
    B. No reserve electrical power in case of an electrical failure in the booster.
    C. Fluid circulation in the steering gear will fail.
    D. Braking power will be reduced with the engine running.   (B3.1)

89. Technician A says that in a school bus air-over-hydraulic brake system, an antifreeze valve may be used. Technician B says that a limiting valve is often used on rear axle brakes. Who is right?
    A. A only
    B. B only
    C. Both A and B
    D. Nether A nor B   (B3.3)

90. School bus wet wheel bearings:
    A. should be repacked every 10,000 miles.
    B. are of a semifloating design.
    C. should be inspected every time a wheel is pulled.
    D. are usually ball bearings.   (C2)

91. When adjusting wheel bearings, Technician A says to torque the adjusting nut to 50 ft.-lbs., then back off the nut one-sixth to one-third turn and install the lock ring. Technician B says wheel end play should be measured with a dial indicator after the adjustment. Who is right?
    A. A only
    B. B only
    C. Both A and B
    D. Neither A nor B   (C3)

92. Which of the following is true when diagnosing hydraulic brake systems?
    A. Chassis vibration during braking can be caused by excessive rotor run out.
    B. Brake grab on one wheel can be caused by an improper brake pedal pushrod adjustment.
    C. Brake drag can be caused by glazed brake linings.
    D. A low spongy pedal can be caused by rotors that are machined too thin.

                                                                    (B1.1)

    chnician A says that it is important not to mix up a left side S-camshaft with a
         ide S-camshaft when performing a brake overhaul. Technician B says that
            king application forces are compounded, S-camshaft splines can be
              o is right?

                                                                    (A2.4)

94. Which of the following procedures is correct when pressure bleeding a hydraulic brake system?
    A. Bleed the right front caliper or wheel cylinder first.
    B. The pressure bleeder is pressurized to 20 to 25 psi.
    C. If equipped, the metering valve must be closed.
    D. Bleed the left rear caliper or wheel cylinder last.   (B1.10)

95. If the floating caliper on school bus hydraulic brakes does not slide freely, which of these conditions will result?
    A. Excessive brake force
    B. Excessive brake pad wear
    C. Reduced braking force
    D. The brakes will grab.   (B1.9)

96. When servicing disc brake assemblies, which of these applies to the pads?
    A. Thickness is indicated by a code on the lining edge.
    B. All semi metallic linings are edge-stamped "FF."
    C. All asbestos-type linings are edge-stamped "EE."
    D. Pads should use the same edge code as original specification.   (B1.9)

97. Technician A says it's OK to machine oversize drums, providing they are routinely inspected when returned to service. Technician B says that turning an oversize drum will sacrifice its strength. Who is right?
    A. A only
    B. B only
    C. Both A and B
    D. Neither A nor B   (B2.2)

98. When testing foot valve leakage, Technician A says that the brake application pressure should be 80 psi. Technician B says that this test should be done every 50,000 miles. Who is correct?
    A. A only
    B. B only
    C. Both A and B
    D. Neither A nor B   (A1.10)

99. All of the following steps in removing a dual circuit, foot application valve are correct **EXCEPT:**
    A. the bus should be on a level surface.
    B. label the brake lines.
    C. mark the valve body in relation to the mounting plate.
    D. maintain brake system pressure.   (A1.10)

100. Technician A says that when brake linings are glazed, increased application pressures are required to stop the vehicle. Technician B says that when linings are glazed on one side of an axle, the result is brake torque imbalance. Who is right?
    A. A only
    B. B only
    C. Both A and B
    D. Neither A nor B   (A2.1)

101. How much free stroke travel should a properly functioning, automatic slack adjuster test at?
    A. Marginal preload
    B. Zero
    C. ½ to ¾ inch
    D. 1 to 1½ inches   (A2.3)

102. A school bus has been parked overnight in winter conditions and on start-up the rear brakes will not release. Which of the following is the LEAST-Likely cause of the condition?
   A. Relay valve piston frozen
   B. Spring brake control valve frozen
   C. Linings frozen to drum
   D. Treadle valve relay piston frozen   (A2.1)

7 Appendices

Answers to the Test Questions for the Sample Test Section 5

1.	A	16.	A	31.	A	46.	D
2.	C	17.	D	32.	B	47.	A
3.	B	18.	C	33.	A	48.	B
4.	A	19.	D	34.	B	49.	B
5.	C	20.	A	35.	A	50.	C
6.	D	21.	C	36.	B	51.	A
7.	C	22.	C	37.	D	52.	A
8.	B	23.	B	38.	B	53.	C
9.	D	24.	D	39.	C	54.	D
10.	A	25.	B	40.	A	55.	B
11.	A	26.	C	41.	D	56.	A
12.	D	27.	A	42.	A	57.	C
13.	D	28.	A	43.	C	58.	D
14.	D	29.	D	44.	B	59.	B
15.	B	30.	C	45.	D	60.	C

Explanations to the Answers for the Sample Test Section 5

Question #1
Answer A is correct. The safest way to cage the power spring in a spring brake chamber is to charge the hold-off chamber with air and then insert the cage bolt into the cage plate and torque. This makes Technician A right. Technician B is wrong because he has failed to understand how a spring brake chamber is plumbed and it is bad practice to remove the hold-off chamber of a spring brake.

Question #2
Answer C is correct because both technicians are right. Technician A is correct in saying that one type of air brake ABS uses a LED display to indicate fault codes. Technician B is also right in saying that it is a requirement of a school bus ABS system to provide full redundancy back-up in the event of an electronic fault: this means that should an ABS electronic fault occur, the system will revert to standard non-ABS operation.

Question #3
Answer B is correct. Slack adjusters, brake shoes, and brake chambers are all considered to part of the foundation brake system-by definition, the components that effect mechanical braking effort at the wheel. Pneumatic lines are used to route air through the air brake system, so they are not considered part of the foundation brake system.

Question #4
Answer A is correct. Technician A is correct in saying that the clevis must be correctly positioned on the brake chamber pushrod. The clevis yoke is threaded onto the brake chamber pushrod, so Technician B is wrong.

Question #5
Answer C is correct because both technicians are right. It is regarded as good practice to change brake fluid when major brake repairs are performed, so Technician A is correct. Any indication of contamination in brake fluid could result in a dangerous condition, so Technician B is also right.

Question #6
Answer D is correct. An air compressor has pistons, a crankshaft, and discharge valve. The governor is always a separate component that may be mounted on the compressor or remotely.

Question #7
Answer C is correct. Restricted airflow, improper brake adjustment, and a ruptured diaphragm can all compromise service application pressure. Grease is a lubricant and when linings become grease saturated, much higher service pressures are required to effect the same retarding effort.

Question #8
Answer B is correct, because the first component to wear out on foundation brake assemblies is usually the brake shoes.

Question #9
Answer D is correct. The slack adjuster should be adjusted so that a 90 degree angle between the slack adjuster and pushrod is achieved when the brakes are fully applied.

Question #10
Answer A is correct. The minimum number of reservoirs on an FMVSS 121 compliant vehicle is three: a supply tank, primary tank, and secondary tank.

Question #11
Answer A is correct. Technician A is right in saying that all current (manufactured after 1998) school buses are required to have automatic slack adjusters. However, Technician B is wrong in saying that automatic slack adjusters increase the allowable free stroke dimension.

Question #12
Answer D is correct. Misaligned actuation linkage, a defective chamber retraction spring, and weak brake shoe springs can all cause sluggish release problems in air brake systems. However, a leaking service chamber will result in low braking force at the wheel but should not affect release timing.

Question #13
Answer D is correct because both technicians are wrong. A pressure differential valve should automatically center after brake service (if this does not happen, the valve is defective), so Technician A is wrong. Because of this, Technician B is also wrong, so his statement about replacing the proportioning valve is also incorrect.

Question #14
Answer D is correct. The clevis yoke is connected to but not part of the slack adjuster.

Question #15
Answer B is correct. The fail relay coil for an ABS system is installed between the power source and modulator assembly, so Technician A is wrong. Technician B is right in saying that the power for the fail relay light is routed through normally closed relay contacts.

Question #16
Answer A is correct. Concave, scored, and threaded drum failures are relatively common. Brinnelling, however, is a bearing failure condition, so this is the exception.

Question #17
Answer D is correct. Brake linings or friction faces are commonly riveted to shoes. Other methods of attaching linings to shoes would be bolts and bonding.

Question #18
Answer C is correct because both technicians are right. Technician A is right in saying that the hydrobooster and balanced steering gear can be integrated into the power steering pump circuit. Technician B is also right because a dedicated hydraulic power source can be used to power a hydrobooster.

Question #19
Answer D is correct because both technicians are wrong. Governor cut-out occurs 20–25 psi above governor cut-in, so Technician A is wrong. Only governor cut-out should be adjusted, so if after setting governor cut-out, cut-in does not occur at 20–25 psi below cut-in, the governor can be assumed to be defective. This makes Technician B wrong too.

Question #20
Answer A is correct. On a full-floating axle, vehicle weight is supported by axle housing; the axle shaft function is only to transmit drive torque to the wheels. Vehicle weight is split between the wheel bearings.

Question #21
Answer C is correct because both technicians are right. Brake lines are replaced with double flare and ISO approved tubing.

Question #22
Answer C is correct. The LEAST-Likely cause of a blowing oil condition is a defective head gasket. Failed oil control rings, intake restrictions, and plugged oil drain-back ports are all common causes of blowing oil conditions in air compressors.

Question #23
Answer B is correct. The purge valve is indicated by the letter C in the figure.

Question #24
Answer D is correct because all the other possible causes directly affect service braking performance. The LEAST-Likely cause of a poor service braking condition would be a leaking hold-off diaphragm; depending on the severity of the leak, this could cause some loading of the power spring onto the actuator, causing brake drag, but not impact on service braking.

Question #25
Answer B is correct. The definition of a balanced braking system is one in which air pressure applied to the service chambers is timed so that the result is the application of mechanical braking force at each wheel at the same moment. This makes answer B correct. If braking forces at each wheel were identical, one axle could be overbraked, the other underbraked. If brake valves were not pneumatically timed, those wheels closest to the application valve would be braked ahead of those farthest away, a dangerous condition.

Question #26
Answer C is correct. A leaking service diaphragm, hose, or relay valve are all capable of causing a massive air loss during service braking while a governor leak would be continuous (not just during service braking) and of considerably less volume.

Question #27
Answer A is correct. Brake free stroke dimension should be measured through an application pressure of at least 80 psi, so Technician A is right. Technician B is wrong because brake free stroke adjustment is defined at the slack adjuster, not by the clevis and pushrod.

Question #28
Answer A is correct. All school bus OEMs recommend that wheel bearings be set with end play and measured with a dial indicator. This makes Technician A right and Technician B definitely wrong; setting a preload on a wheel bearing can result in a dangerous wheel-off condition.

Question #29
Answer D is correct. Of the conditions listed, short piston pad life could only be caused by a caliper assembly that does not slide, meaning that all braking force is applied directly from the piston, none by leverage to the opposing side.

Question #30
Answer C is correct because both technicians are right. Technician A is correct when he says that an ABS produces an AC electrical signal. Technician B is also correct in describing the redundancy (back-up) required when ABS is disabled, meaning that the system must revert to standard, non-ABS operation.

Question #31
Answer A is correct. Wear between the clevis pin and yoke causes mechanical play, which results in overstroking, so Technician A is correct. Technician B is wrong because automatic slack adjusters, even when they are functioning properly, do require frequent inspection and occasional adjustment especially when a vehicle is operated over rough terrain.

Question #32
Answer B is correct. Hygroscopic describes the tendency to absorb moisture from atmospheric air that is a characteristic of many brake fluids. However, not all heavy-duty brake fluids are hygroscopic, so Technician A is wrong. Technician B is right; he correctly defines the term hygroscopic.

Question #33

Answer A is correct. All the answers relate to the practice of removing a piggyback assembly on a spring brake except for removing the hold-off chamber clamps. Current practice is to NEVER separate the hold-off chamber for any reason.

Question #34

Answer B is correct. An aftercooler type air dryer functions by cooling the hot, moisture laden air delivered to it from the compressor; when this compressed air is cooled, moisture condenses to liquid form and is separated and discharged by the air dryer.

Question #35

Answer A is correct. A quick-release ratio valve is used to modulate brake application pressure to the steering axle wheels on a school bus, so it is located close to the front service chambers.

Question #36

Answer B is correct. Technician A is incorrect because it is illegal to machine a brake drum measured to be oversize. Technician B is correct; drums do distort in storage, so it is good practice to machine them prior to installation.

Question #37

Answer D is correct. The figure shows a typical hydraulic power booster circuit in which the function of the power booster is to assist pedal braking effort. However, the power booster hydraulic circuit and the actual brake hydraulic circuit are quite separate, so the actual force used to effect braking is hydraulic from the master cylinder.

Question #38

Answer B is correct. Bus foundation brakes are mounted to a spider, which is itself bolted to the axle flange.

Question #39

Answer C is correct. When the ABS fails to receive a within range signal from a wheel speed sensor, it shuts down ABS management of only that portion of the circuit. However, due to the system redundancy requirement, the foundation brakes will continue to function at 100 percent potential.

Question #40

Answer A is correct. Technician A correctly describes the critical length of a slack adjuster; this defines the leverage of the slack adjuster used to actually determine brake torque developed at the S-cam. Technician B is wrong because if the slack adjuster critical length is increased, the effect will be to produce unbalanced braking, which could be dangerous.

Question #41

Answer D is correct. The consequence of a broken retraction spring in a foundation brake assembly is the brakes will drag because the function of this spring is to pull the shoes away from the drum after a brake application.

Question #42

Answer A is correct. Worn S-cam bushings, a broken anchor spring, and grease on the cam profile could all be the cause of an S-cam rollover. The Least-Likely cause of an S-cam rollover is an out-of-adjustment brake, as this would mean less available brake torque to the foundation assembly.

Question #43

Answer C is correct because both technicians are right. Technician A is correct in saying that no attempt should be made to repair any part of the piggyback assembly; current practice dictates that the assembly be removed and replaced. Technician B is also correct in saying that it is a legal requirement that the power spring in any spring brake assembly be disarmed before the unit is disposed of. This process means separating the hold-off chamber in a disarmament chamber, thus releasing the power spring.

Question #44

Answer B is correct. In a dual circuit foot valve, the relay piston responsible for actuating the secondary circuit is itself actuated pneumatically by the primary section of the valve when the valve is functioning properly.

Question #45

Answer D is correct. The relay piston is actuated mechanically by the primary piston in the event of loss of primary circuit air. The foot pedal acts on the primary piston, which drives it down until it contacts the relay piston to actuate the secondary circuit.

Question #46

Answer D is correct. The figure shows a standard dash park control valve, which controls air supplied to the spring brake hold-off circuit. Pulling it out exhausts air from the hold-off circuit placing the vehicle in park mode.

Question #47

Answer A is correct. Technician A is correct in saying that when disk brakes are not properly adjusted, mechanical advantage is reduced. Disk brakes produce superior braking to S-cam brakes, but this depends on the brakes being properly adjusted. Technician B is incorrect; the slack adjuster and pushrod should form a 90-degree angle when the brakes are fully applied, not released.

Question #48

Answer B is correct. Only S-cam-actuated brakes are a drum brake system so Technician A is wrong and Technician B is right.

Question #49

Answer B is correct. There is no such thing as an anticompounding valve on a front disc-rear drum system, so Technician A is wrong. Technician B is right; a proportioning valve is used in a non-ABS disc-drum brake system and may also be used in an ABS system to ensure system redundancy.

Question #50

Answer C is correct because both technicians are right. When a hold-off diaphragm leaks, the result can be insufficient air pressure to prevent the power spring from loading the chamber pushrod applying brake pressure. So Technician A is right. Technician B is also right; a broken or weak return spring can cause brake drag.

Question #51

Answer A is correct. A blocked or restricted compensating port in a master cylinder can cause brake drag, so Technician A is right. However, Technician B is wrong in saying that excessive free play could cause brake drag, though you should note that insufficient free play at the pedal can cause brake drag.

Question #52

Answer A is correct. If a master cylinder reservoir cover diaphragm is not installed, the brake fluid will become contaminated.

Question #53

Answer C is correct because both technicians are right. Technician A is right because a sticking caliper piston can cause brake drag. Technician B is also right; rotors machined below specified minimum thickness will be unable to sustain the heat they are subjected to, and the result will be warpage.

Question #54

Answer D is correct. This question explores your knowledge of four brake drum failure conditions. Bell-mouthing occurs when inboard drum wear is greater than outboard drum wear. Concave wear occurs when wear at the center of the drum is greater than at the edges. Out of round means that the drum is eccentric rather than circular. A drum that indicates greater wear at the edges than at the center is convex.

Question #55
Answer B is correct. Technician A is wrong; low brake fluid level could cause loss of braking but not brake fade, which is generally associated with overheating of the foundation brakes. Technician B correctly describes one of the reasons for brake fade. Drums that are worn or oversize have been subjected to more heat than they are able to dissipate, resulting in lowering the coefficient of friction or aggressiveness of the brakes.

Question #56
Answer A is correct. Technician A is right in saying that to properly inspect the brake pads on a bus with hydraulic disc brakes, the wheels should be removed. Technician B is wrong; if lining to rotor clearance was to be measured immediately after an application, clearance would be more likely less than greater than the specification.

Question #57
Answer C is correct because both technicians are right. A restricted one-way check valve in the vacuum hose to a hydrovac booster could result in excessive pedal effort, so Technician A is right. A sticking power piston in the hydrovac could also cause this problem, so Technician B is also right.

Question #58
Answer D is correct. Some air-over-hydraulic parking systems still use cable-actuated parking brakes.

Question #59
Answer B is correct. A slack adjuster is connected to an S-camshaft by means of splines on the camshaft.

Question #60
Answer C is correct. Brake applied stroke dimension should measure between 1.00″ and 1.50″ at the slack adjuster.

Answers to the Test Questions
for the Additional Test Questions Section 6

1.	B	27.	B	53.	B	79.	C
2.	D	28.	B	54.	A	80.	B
3.	C	29.	A	55.	D	81.	A
4.	A	30.	A	56.	C	82.	C
5.	D	31.	B	57.	C	83.	B
6.	C	32.	C	58.	C	84.	B
7.	C	33.	C	59.	A	85.	C
8.	C	34.	B	60.	C	86.	C
9.	B	35.	B	61.	D	87.	B
10.	A	36.	C	62.	D	88.	A
11.	C	37.	A	63.	B	89.	A
12.	D	38.	A	64.	A	90.	C
13.	D	39.	D	65.	B	91.	C
14.	C	40.	D	66.	B	92.	A
15.	B	41.	A	67.	D	93.	C
16.	C	42.	C	68.	A	94.	B
17.	C	43.	B	69.	D	95.	C
18.	D	44.	B	70.	A	96.	D
19.	C	45.	C	71.	D	97.	B
20.	A	46.	C	72.	C	98.	A
21.	C	47.	A	73.	C	99.	D
22.	D	48.	C	74.	A	100.	C
23.	B	49.	C	75.	A	101.	C
24.	C	50.	B	76.	A	102.	D
25.	C	51.	C	77.	B		
26.	D	52.	D	78.	C		

Explanations to the Answers for the Additional Test Questions Section 6

Question #1
Answer B is correct. A two-way check valve is designed to receive pressure from two sources and will bias-shuttle to deliver the higher source pressure to its single outlet. This means that if one of the two air circuits (primary and secondary) was to lose pressure, air from the functioning circuit can be routed into the still functioning circuit.

Question #2
Answer D is correct. Because brakes have to be specified for the maximum braking force required when a fully loaded vehicle is panic braked, brake systems are engineered for much higher braking forces than are required for everyday stops and starts. The majority of air brake applications are made with an application of 20 psi or less.

Question #3
Answer C is correct because both technicians are right. Technician A is correct in describing the park/emergency brake system as being separate and distinct from the service brake air circuit. Technician B is also correct in saying that the hold-off chambers produce pneumatic force in the opposite direction to the service chambers, because their function is to hold-off rather than apply.

Question #4
Answer A is correct. When grease lubricated wheel bearings are used, it is required that a good quality, high temperature grease be used to repack them, so Technician A is right. Technician B is wrong; bearings and their races must remain matched. That means they cannot be switched, and when they are replaced, both the roller bearing and its race should be replaced together.

Question #5
Answer D is correct. Brake compounding occurs when mechanical force applied by the power spring and pneumatic force applied by the service chamber combine or "compound." This can result in an actuation pressure that exceeds the specified maximum by 150 percent, which can damage foundation brake components. Another way of saying this is the simultaneous application of spring and service brake forces acting on the foundation brakes.

Question #6
Answer C is correct because both technicians are right. Technician A correctly describes how to install a seal onto the worm gear of a slack adjuster, so he is right. It is also important not to strike a seal once it has bottomed into the seal bore, so Technician B is also right.

Question #7
Answer C is correct. Hydraulic brake boosters can be used on hydraulic and air-over-hydraulic brake systems, so both technicians are right.

Question #8
Answer C is correct because both technicians are right. S-cam-actuated foundation brake systems are the most common type of foundation brake system found on air brake equipped school buses, so Technician A is right. Slack adjusters working with S-cams convert the linear force developed by the brake chamber into brake torque, so Technician B is also right.

Question #9
Answer B is correct. The discharge valve in a typical air compressor is a one-way check valve that permits compressed air to exit the cylinder, but closes when cylinder pressure is less than pressure from the other side of the discharge valve.

Question #10
Answer A is correct. The pressure differential valve in a hydraulic brake system is a pressure warning valve that is balanced between primary and secondary circuit pressure. In the event of a pressure imbalance, the valve would bias toward the low circuit and illuminate a brake warning light.

Question #11
Answer C is correct because both technicians are right. Pulse cycle rates in current bus ABS systems peak at about 7 times per second, so Technician A is right. Technician B is also correct; pulse cycle rates in current hydraulic brake systems peak at about 20 times per second.

Question #12
Answer D is correct. Brake friction facings are attached to shoes by bonding, rivets, and screws and nuts, but never by welding.

Question #13
Answer D is correct since the figure shows three types of relay valve.

Question #14
Answer C is correct because air is present in both the service and hold-off chambers. Because the vehicle is moving, it indicates the parking brakes are released, meaning that there is air in the hold-off chambers. When the driver makes a service application, air is now charged to the service chambers.

Question #15
Answer B is correct. In any S-cam brake circuit, the role of the slack adjuster is that of a lever and its function is to multiply brake torque.

Question #16
Answer C is correct because both technicians are right. Brake systems function by retarding vehicle movement by converting the energy of motion to friction, which is then dissipated as heat. Therefore, both technicians are right.

Question #17
Answer C is correct because both technicians are right. Brake timing is affected by the size of lines. Air is highly compressible, so the larger the volume of compressed air in an air line, the higher the compressibility factor. Technician A is right. Technician B is also right because the flow area defined by any fitting or line will impact on brake timing.

Question #18
Answer D is correct since both technicians are wrong. Normal testing of a master cylinder does not require the use of a vacuum gauge, so Technician A is wrong. Wheel cylinders must be inspected during a complete brake overhaul but replaced only if there is evidence of wear or failure. So Technician B is also wrong.

Question #19
Answer C is correct because both technicians are right. Technician A is correct; the angle between the slack adjuster and the chamber pushrod should be 90 degrees when the brakes are fully applied. Technician B is also right; the optimum adjustment performed using the wheel-up method would ensure the slack adjuster had the shortest possible stroke with zero drag.

Question #20
Answer A is correct. Overloading is the LEAST-Likely cause of a wheel bearing failure because bearings are specified for the expected load capabilities of a vehicle with a generous margin of error. Contamination, damaged race, and improper lubricant are more common causes of bearing failure.

Question #21
Answer C is correct. There are 100 teeth on the reluctor wheel used in a typical wheel speed sensor equivalent to 3.6 degrees per tooth.

Question #22
Answer D is correct because hydraulic brake systems should be cleaned with alcohol. Water, gasoline, and mineral spirits will either contaminate a hydraulic brake system or damage components by swelling seals.

Question #23
Answer B is correct. Technician A has not identified the cause of the problem and, because the air pressure is not building to a value exceeding 150 psi, it indicates that the safety pop-off valve is properly functioning. Technician B is correct; the first step in troubleshooting this condition should be to remove the unloader line and check for the presence of signal or stuck unloader assembly.

Question #24
Answer C is correct. A brake adjustment is being performed by rotating the star wheel.

Question #25
Answer C is correct because both technicians are right. Technician A is right; the function of the electric pump is emergency back-up in the event of hydraulic supply failure to the hydromax. Technician B is also right; unassisted braking is possible in the event of hydromax failure but it would be difficult.

Question #26
Answer D is correct. An ABS system described as being 4S/4M (4 wheel sensor inputs/4 modulators) means that all four wheels are modulated.

Question #27
Answer B is correct. Technician A is wrong because rotor thickness is checked with a micrometer. Technician B is right; rotor run out is being indicated with a dial indicator.

Question #28
Answer B is correct. When a gear driven air compressor goes into unloaded cycle, an unloader signal from the governor actuates the unloader in the compressor and holds the inlet valves open. The compressor is still rotated, but because the inlet valves are open, it simply sucks and blows air through the inlet valves.

Question #29
Answer A is correct. Technician A is right. Although the drum is just within the legal maximum limit, it exceeds the machine limit, so it cannot be turned. It would simply not be good economic practice to return this drum to service. Technician B is wrong; it is not permitted to turn a drum that has exceeded its machine limit.

Question #30
Answer A is correct. In the figure, a single tank containing a supply reservoir and secondary reservoir is shown; the two reservoirs are divided by a baffle. A one-way check valve is located in the baffle, which permits supply air to pass into the secondary reservoir but not vice versa.

Question #31
Answer B is correct. Oil saturated or glazed lining should be replaced, not cleaned, so Technician A is wrong. Technician B is right; it is good practice to replace the linings on the brake along with those on the other side of the axle to ensure good brake balance.

Question #32
Answer C is correct. Brake fluid should be stored in a clean, sealed container (glass is OK) that has as little air in it as possible. Under these conditions, brake fluid should have a good shelf life. Contact with contaminants such as other shop liquids can adversely affect the life and performance of brake fluid.

Question #33
Answer C is correct. The component shown in the figure is the service retraction spring.

Question #34
Answer B is correct. Technician A is wrong; the standard ABS sensor used on a vehicle ABS system is an AC pulse generator, so the voltage produced is AC. Technician B is right; the ABS-ECU does read the AC frequency produced by the pulse generator and from that calculates wheel speed.

Question #35
Answer B is correct. The function of the steel tube inserted into the front of the crankshaft on an air compressor is to supply engine lubricant to the compressor lubrication circuit. Technician A is wrong, and Technician B is right; the compressor will indeed rapidly fail if this tube is not installed.

Question #36
Answer C is correct because both technicians are right. Hydraulic disc brake calipers can have one, two, and four pistons.

Question #37
Answer A is correct. Hydraulic lines used on brake systems must conform to SAE J189 standards, so Technician A is right. Technician B is wrong because a pressure bleeder should not be used to bleed the hydro-boost circuit.

Question #38
Answer A is correct. Both front and rear wheel bearings are adjustable, so Technician A is right. Technician B is wrong; a brass drift should never be used to remove a wheel bearing race because the minute cutting produced can be difficult to remove from the hub assembly and can cause rapid bearing damage.

Question #39
Answer D is correct. Hydraulic single piston disc brakes operate on an action and reaction principle.

Question #40
Answer D is correct. Master cylinder pistons are retracted mechanically by return springs.

Question #41
Answer A is correct. Technician A is correct; S-cam-actuated brakes are released after an application by the foundation assembly return springs and by the service chamber retraction springs. Technician B is wrong; the clevis yoke is threaded to the chamber pushrod and connected to the slack adjuster by means of a clevis pin.

Question #42
Answer C is correct because both technicians are right. Technician A is right because vented rotors are sometimes engineered to have a thicker inboard pad than the outboard pad. And Technician B is also right because solid disc systems usually are specified with pads of equal thickness.

Question #43
Answer B is correct. Technician A is wrong because there is no check valve on a governor. Technician B is right; too much clearance at the compressor unloading valves could cause system pressures to rise above the intended cut-out pressure.

Question #44
Answer B is correct. During service breaking, braking force is delivered by service application air.

Question #45
Answer C is correct because both technicians are right. Technician A is correct because unitized wheel seals should be replaced any time a wheel is pulled. Technician B is also correct because removing a wheel seal with a pair of heal bars is better than pounding them out with a drift.

Question #46
Answer C is correct because both technicians are right. Technician A is right because the dash low air warning light should illuminate at or slightly above 60 psi. Technician B is also right. If it takes 60 psi air pressure to release spring brakes, when hold-off pressure drops below that value, the power spring begins to exert pressure on the chamber pushrod, applying the brakes.

Question #47
Answer A is correct. It is good practice to replace all the springs in the foundation assembly anytime a brake job is performed on a school bus because springs lose their tension when continually heated and cooled. Technician A is right. Technician B is wrong; the maximum permissible out-of-round specification on a drum brake system is likely to be between 0.008 inches and 0.015 inches.

Question #48
Answer C is correct because both technicians are right. The figure shows a typical foot application valve. Technician A is right when he says that the valve is actuated mechanically in both normal and failure modes, specifically it is actuated by driver foot pressure. Technician B is also right; when a complete loss of primary air pressure occurs, the primary piston is stroked downward by foot pressure until it mechanically contacts the relay piston, actuating the secondary circuit.

Question #49
Answer C is correct because both technicians are right. Technician A is correct because compressor discharge air pressures often reach 300° F. Technician B is also right because most school bus compressors are incorporated into the engine cooling circuit.

Question #50
Answer B is correct. School bus disc wheel assemblies do not combine the wheel and hub into a single unit, so Technician A is wrong. Technician B is right; the wheel bearing nut can be locked into position by a lock nut, split forged nut, or castellated nut and cotter pin.

Question #51
Answer C is correct because both technicians are right. In discussing foot application valve operation, Technician A is right; air pressure at the delivery port(s) should be in proportion to treadle mechanical travel. Technician B is also right; a 1-inch air bubble produced in 3 seconds during full application is consistent with the maximum leakage specification.

Question #52
Answer D is correct because both technicians are wrong. Master cylinder sub-components should never be washed in anything but alcohol, so Technician A is wrong. Technician B is also wrong because an aluminum master cylinder bore is surface hardened and should never be honed.

Question #53
Answer B is correct. Technician A is wrong; leakage greater than a 1-inch bubble over 3 seconds at the exhaust during full application pressure is excessive. Technician B is right; when service application pressure is released, the front brake chambers should exhaust immediately.

Question #54
Answer A is correct. Of the reasons listed, only worn compressor rings could account for slow buildup in the supply system.

Question #55
Answer D is correct. An automatic reservoir drain valve (spitter valve) spits when sump pressure exceeds reservoir pressure by about 2 psi.

Question #56
Answer C is correct. When any pulley belt has worn to the extent that it bottoms into the pulley groove, it should be replaced.

Question #57
Answer C is correct because both technicians are right. Technician A is correct; a single-cylinder compressor can unbalance an engine if it is not timed to the engine accessory drive (check the OEM instructions). Technician B is also correct; most two-cylinder compressors are balanced units and do not require timing when coupled to the engine accessory drive.

Question #58
Answer C is correct because both technicians are right. Technician A is describing correct procedure to be observed when installing a wheel assembly to an axle. Technician B is also correct; most school bus drive axle wheel ends are lubricated by oil from the differential carrier through the axle housing.

Question #59
Answer A is correct. Rotating the adjusting screw counterclockwise increases the governor cut-out pressure.

Question #60
Answer C is correct because both technicians are right. Technician A is right; daily draining of air reservoirs not equipped with spitter valves is recommended practice. Technician B is also right; automatic drain valves should be periodically checked. It should be noted that this periodic check should involve completely draining the tank.

Question #61
Answer D is correct. The LEAST-Likely cause of water in an air reservoir is continuous operation of the system in humid conditions because the system, when functioning properly, is designed to handle conditions of high humidity.

Question #62
Answer D is correct because only a severely out-of-adjustment brake would allow this to happen. Under the circumstances, there would be no air in the hold-off chamber, so the power spring would be fully applied to the actuator pushrod. Even a broken power spring or worn brake linings would not allow the wheel to be rotated by hand.

Question #63
Answer B is correct. You are looking for the incorrect procedure or statement in this question-that is, signal pressure should equal inlet pressure. Signal pressure is the service application pressure delivered from the application valve, whereas inlet pressure should be close to system pressure.

Question #64
Answer A is correct. All the statements are false except that stating the air lines connect the delivery ports to the service brake chambers.

Question #65
Answer B is correct. Technician A is wrong; the locking collar on a manual slack adjuster should not be staked to the adjusting nut. Technician B is correct; if the lock collar will not retract to engage, the slack adjuster should be replaced.

Question #66

Answer B is correct. Technician A is wrong. The only adjustment on an air pressure governor is the cut-out pressure, so if the cut-out pressure is set correctly and the cut-in pressure is out of specification, the governor should be replaced. Technician B is right; the difference between governor cut-in and cut-out should be between 20 and 25 psi.

Question #67

Answer D is correct. Because the governor cut-out is set correctly and cut-in is out of specification, the governor has to be changed to prevent frequent cycling of the compressor.

Question #68

Answer A is correct. Technician A is correct; some air brake systems feed the secondary reservoir from the primary reservoir (this cannot happen the other way around), though current practice has the primary and secondary reservoirs fed separately and directly from the supply tank. Technician B is wrong; a two-way check valve routes the higher of two source pressures to the valve outlet.

Question #69

Answer D is correct because both technicians are wrong. When the low air pressure warning fails to extinguish after startup, it would mean that system pressure was not higher than 60 psi in either or both of the onboard circuits. After startup, the operator would have to attempt to build system pressure before making a judgment on whether there was a malfunction, so Technician A is wrong. Technician B is wrong in saying that there is a problem with the operation of the system, although a short circuit could cause the low-pressure warning light to stay on even if system pressure was above 60 psi.

Question #70

Answer A is correct. Technician A is right; if a low-pressure warning light does not illuminate when system pressure drops below 60 psi, the problem could be as simple as a defective bulb. Technician B is wrong in saying a defective dash pressure gauge could cause this problem because the low-pressure warning light circuit must be separate from the dash pressure gauge circuit.

Question #71

Answer D is correct. All the answers relate to brake torque imbalance problems except for a vehicle overload condition. While overload might tax brake system potential, it would not in itself cause brake torque imbalance.

Question #72

Answer C is correct because both technicians are right. Technician A is right; the conditions described could cause S-cam rollover. Technician B is also right; these conditions could cause loss of steering control under braking.

Question #73

Answer C is correct. This question is asking you to identify the incorrect statement, and the cam profile should never be lubricated with anything.

Question #74

Answer A is correct. There is no cam actuator used on an air brake system.

Question #75

Answer A is correct. Technician A is right; the same fastener tightening sequence is used with both bolted and riveted linings. Technician B is wrong; his idea of greasing the actuator roller profiles is not good practice and could result in cam rollover.

Question #76

Answer A is correct. Severe heat checking on brake drums can cause accelerated liner wear, so Technician A is right. Technician B is wrong again; machine limits on bus brake drums are typically 0.080 to 0.090 inch, never as high as 0.130 inch.

Question #77
Answer B is correct because pedal pulsation is the more likely condition. Extreme rotor run out in a hydraulic brake system is Most-Likely to cause pedal pulsation, but you should note that under panic braking, it could provide directional control problems.

Question #78
Answer C is correct. Worn foundation brake linings would be the LEAST-Likely cause of a parking brake that would not hold.

Question #79
Answer C is correct. Current practice is to replace either the spring brake assembly or install a piggyback unit if the hold-off diaphragm fails, mainly because it can be dangerous separating a spring brake at the hold-off chamber.

Question #80
Answer B is correct. All the reasons listed as answers could cause an ABS operational fault, except for leaking hydraulic lines, which would cause a brake system fault rather than an ABS fault.

Question #81
Answer A is correct. ATC (automatic traction control) uses the ABS hardware to prevent spinout conditions in bus drive wheels. When one drive wheel velocity exceeds the mean velocity of the other, light brake pressure is applied to the spinning wheel to more evenly divide torque distribution at the differential and cancel the spinout. Technician A accurately describes ATC operation, but Technician B is wrong because ATC only functions on drive axle wheels, and on a typical school bus, that means on two of the four wheels.

Question #82
Answer C is correct. Both two and four channel ABS systems are used on school buses, but four channel is most common.

Question #83
Answer B is correct. Brake fade is caused by heated foundation brakes in both hydraulic and air brake systems and when drums are machined beyond limit. The mass of material remaining is often insufficient to dissipate heat sufficiently quickly, so overheated brakes result.

Question #84
Answer B is correct. Hydraulic brake components should always be cleaned with alcohol and then completely dried. The residues left after cleaning with soap, gasoline, and mineral spirits can damage rubber seals and break down the hydraulic fluid.

Question #85
Answer C is correct. Disc brakes have much lower lag times than shoe brakes, so on a front disc-rear drum system, a metering valve is used to delay the application of the front brakes-that is, time the brakes so that front and rear are applied at the same time.

Question #86
Answer C is correct because both technicians are right. Technician A is right because, when a brake system with a metering valve is bled using a pressure bleeder, the metering valve must be held open. Technician B is also right; when the same system is bled manually, the metering valve does not have to be held open.

Question #87
Answer B is correct. A brake drum worn more at the center than at the edges is described as being concave.

Question #88
Answer A is correct. If the reserve electric motor for a hydraulic power brake system failed, no reserve stops without the engine running would be possible.

Question #89
Answer A is correct. Technician A is correct because an antifreeze valve can be incorporated into an air-over-hydraulic brake system. Technician B is wrong because a limiting valve is used over the front axle brakes, not the rear.

Question #90
Answer C is correct. Wet bearings used on busses are of a taper roller design and should not be packed with grease as it compromises the superior lubrication provided by liquid gear oils. It is important to inspect bearings every time a wheel assembly is pulled.

Question #91
Answer C is correct because both technicians are right. Technicians A and B are accurately describing steps in one OEM wheel end adjustment procedure; always use the OEM recommended wheel end adjustment.

Question #92
Answer A is correct. Chassis vibration during braking can be caused by excessive rotor run out, whereas with the other answer options, the cause does not correlate to the condition.

Question #93
Answer C is correct because both technicians are right. Technician A is correct; left and right side S-camshafts should not be mixed. Technician B is also right; compounding of brake application forces can result in 150 percent specified force applied to the S-cam shaft and the result can be bent or broken splines.

Question #94
Answer B is correct. The procedures listed are all incorrect except for setting the pressure bleeder pressure at 20–25 psi.

Question #95
Answer C is correct. If a floating caliper does not freely slide, the result will be reduced braking force.

Question #96
Answer D is correct. When replacing disc brake pads, the edge codes on the new friction facings should correspond to the original specification edge code.

Question #97
Answer B is correct. Technician A is wrong; it is never OK and is actually illegal to machine oversized drums. Technician B is right; oversized drums lose strength and overheat more rapidly.

Question #98
Answer A is correct. Foot valve exhaust leakage should be tested when the valve is producing an application pressure of 80 psi or higher, so Technician A is right. Technician B is wrong; testing of a foot valve is performed when malfunction is suspected or at a major PM inspection.

Question #99
Answer D is correct. When a dual circuit, foot application valve is changed, the air system should be drained entirely, so maintaining brake system pressure is not a requirement of the procedure.

Question #100
Answer C is correct because both technicians are right. Technician A is right in saying that when brake linings become glazed, increased application pressures are required to stop the vehicle. Technician B is also right; when brake linings on one side of an axle are glazed the result is brake torque imbalance.

Question #101
Answer C is correct. The correct free stroke dimension for most slack adjusters should be between ½ and ¾ inches.

Question #102
Answer D is correct. Foundation brake freeze-ups are not uncommon in school buses during the winter because they tend to be parked overnight. All of the conditions should cause a no-release problem, but the LEAST-Likely is a frozen foot valve relay piston.

Glossary

ABS Acronym for antilock brake system.

Absolute Pressure The zero point from which pressure is measured.

Actuator A device that delivers motion in response to an electrical signal.

Additive An additive intended to improve a certain characteristic of the material.

Air Bag An air-filled device that functions as the spring on axles that utilize air pressure in the suspension system.

Air Brakes A braking system that uses air pressure to actuate the brakes by means of diaphragms, wedges, or cams.

Air Compressor An engine-driven pump supplying high pressure air to a bus brake system.

Air Dryer A unit that removes moisture. Located between the compressor and supply tank.

Air Hose Any flexible air line, such as one from a relay valve to a brake chamber that supplies air.

Air-Over-Hydraulic Brakes A brake system using a hydraulic circuit actuated by air pressure.

Air-Over-Hydraulic Intensifier A device that changes the pneumatic air pressure from the treadle brake valve into hydraulic pressure at the wheel cylinders.

Air Spring An air-filled bag that functions as the spring on axles that use air pressure suspensions.

Air Spring Suspension Single or multiaxle suspension using air bags for springs.

Air Timing Time required for air to be transmitted to or released from each brake, starting the instant the driver moves the brake pedal.

Ambient Temperature Temperature of the surrounding or prevailing air. Normally, it is considered to be the temperature in the service area where testing is taking place.

Amp (A) Acronym for ampere.

Ampere (A) The unit for measuring electrical current.

Analog Signal A voltage signal that varies within a given range (from high to low, including all points in between).

Analog-to Digital Converter (A/D converter) A device that converts analog voltage signals to a digital format; located in a section of the ECM called the input signal conditioner.

Analog Volt/Ohmmeter (AVOM) A test meter used for checking voltage and resistance. Analog meters should not be used on solid state circuits.

Antilock Brake System (ABS) A computer-controlled brake system having a series of sensing devices at each wheel that control braking action to prevent wheel lock-up.

Antilock Relay Valve (ARV) In an antilock brake system, the device that usually replaces the standard relay valve used to control rear axle service brakes and perform relay valve functions.

Application Valve A foot- or hand-operated brake valve that modulates air pressure to the service chambers.

Articulation Vertical movement of the front driving or rear axle relative to the frame of the vehicle to which they are attached.

ASE Acronym for Automotive Service Excellence, a trademark of National Institute for Automotive Service Excellence.

Aspect Ratio A tire term calculated by dividing the tire's section height by its section width.

Atmosperic Pressure The weight of the air at sea level; 14.696 pounds per square inch (psi) or 101.33 kilopascals (kPa).

ATC Automatic traction control. Used in conjunction with ABS on some school buses.

Automatic Slack Adjuster Device that automatically adjusts clearance between the brake linings and the brake drum or rotor. The slack adjuster controls clearance by sensing stroke length of the pushrod for the air brake chamber.

Axis of Rotation The center line around which a gear or part revolves.

Axle Seat A suspension component used to support and locate the spring on an axle.

Backing Plate A metal plate that serves to keep dirt away from the brake shoes and other drum brake hardware.

Bellows A movable cover or seal that is pleated or folded like an accordion to allow for expansion and contraction.

Bias Ply A tire in which belts and plies are laid diagonally or crisscrossing each other.

Bleed Air Tanks The process of draining condensation from air tanks to increase air capacity and brake efficiency.

Block Diagnosis Chart A troubleshooting chart that lists symptoms, possible causes, and probable remedies in columns.

Boss The race of a bearing or a pivot bore.

Brake Control Valve A valve that modulates air from the service reservoirs to service lines and brake chambers.

Brake Disc Steel disc used in a braking system with a caliper and pads. When the brakes are applied, pads on each side of a rotor are forced against the disc, imparting braking force. This type of brake is resistant to brake fade.

Brake Drum A cast metal bell-like cylinder attached to the wheel used to house brake shoes and provide a friction surface for stopping a vehicle.

Brake Fade Condition that occurs when friction surfaces become hot enough to cause the coefficient of friction to drop.

Brake Lining Friction material used to line brake shoes or brake pads. Withstands high temperatures and pressure. The molded material is either riveted, bolted, or bonded to the brake shoe.

Brake Pad Friction lining and plate assembly that is forced against the rotor to effect braking action in a disc brake system.

Brake Shoe Arced metal component, faced with brake lining, which is forced against the brake drum to effect braking action.

Brake Shoe Rollers Hardware that attaches to the web of the brake shoes by means of roller retainers. The rollers ride on the S-cam profiles.

Brake System Vehicle system that slows or stops a vehicle. A combination of foundation brakes and a control system.

Bump Steer Erratic steering caused from rolling over bumps, cornering, or heavy braking. Same as orbital steer and roll steer.

Caliper Disc brake component that changes hydraulic or air pressure into mechanical force to press brake pads against the rotor and stop the vehicle. Calipers come in three types: fixed, floating, and sliding, and can have one or more pistons.

Camber The attitude of a wheel and tire assembly when viewed from the front of a vehicle. If it leans outward, away from the vehicle at the top, the wheel is said to have positive camber. If it leans inward, it is said to have negative camber.

Cam Brakes Brakes that force shoes into the brake drum using an S-shaped cam.

Caster The angle formed between the kingpin axis and a vertical axis as viewed from the side of the vehicle. Caster is considered positive when the top of the kingpin axis is behind the vertical axis.

Center of Gravity The point around which the weight of a vehicle is evenly distributed; the point of balance.

Check Valve A valve that allows air to flow in one direction only. It is a federal requirement to have a check valve between the wet and dry air tanks.

Circuit Complete path of electrical current flow including the generating device. When the path is unbroken, the circuit is closed and current flows. When the circuit continuity is broken, the circuit is open and current flow stops.

COE Acronym for cab-over-engine.

Coefficient of Friction Measurement of the friction developed between two objects in contact when one of the objects is drawn across the other.

Compression Applying pressure to a spring causing it to reduce its length in the direction of the compressional force.

Compressor A mechanical device that increases pressure within a container by pumping air into it.

Condensation The process by which gas changes to a liquid.

Conductor Any material that permits the electrical current to flow.

Controlled Traction A type of differential that uses a friction plate assembly to transfer drive torque from the vehicle's slipping wheel to the one wheel that has good traction or surface bite.

Cycling Repeated on-off action of the air conditioner compressor.

Dash Control Valve Handoperated valves located on the dash. It includes the parking system valve.

Data Bus Circuits through which computers communicate with other electronic devices.

Deadline To take a vehicle out of service.

Deburring To remove sharp edges from a cut.

DER Acronym for Department of Environmental Resources.

Diagnostic Flow Chart A chart that provides a systematic approach to component troubleshooting and repair. Found in service manuals and are model specific.

Dail Calipers A measuring instrument capable of taking inside, outside, depth, and step measurements.

Digital Binary Signal A signal that has only two values; on and off.

Digital Volt/Ohmmeter (DVOM) A type of test meter recommended by most manufacturers for use on solid state circuits. Usually referred to as a DMM.

Disc Brake A steel rotor used in a braking system with a caliper and pads. When the brakes are applied, the pad on each side of the rotor is forced into the rotor, imparting retarding force. This type of brake is resistant to brake fade.

Dispatch Sheet A form used to keep track of dates when the work is to be completed. Some dispatch sheets follow the job through each step of the servicing process.

DMM Digital multimeter. A digital volt, ohm, and ammeter recommended for diagnosing electronic circuits such as ABS.

Dog Tracking Off-center-line tracking of the rear wheels with the front.

DOT Acronym for Department of Transportation.

Drum Brake Brake system in which friction is created by shoes pressing against a rotating drum.

Dual Hydraulic Braking System Brake system consisting of a tandem, master cylinder, essentially two master cylinders formed by two separate pistons and fluid reservoirs in a single cylinder.

ECU Acronym for electronic control unit.

EPA Acronym for the Environmental Protection Agency.

Exhaust Brake A valve in the exhaust pipe between the manifold and the muffler. A slide mechanism which restricts the exhaust flow, causing exhaust back pressure to build up in the engine's cylinders and transform the role of the engine into an energy absorbing compressor driven by the drive wheels.

False Brinelling The polishing of a surface that is not damaged.

Fanning the Brakes Applying and releasing the brakes in rapid succession on a long downgrade.

Fatigue Failures The progressive destruction of a shaft or gear teeth material usually caused by overloading.

Fault Code A code that is recorded into computer memory.

Federal Motor Vehicle Safety Standard (FMVSS) Federal standards that govern system safety standards in the U.S.

Federal Motor Vehicle Safety Standard No:121 (FMVSS 121) Federal standard that covers air brake systems.

FHWA Acronym for Federal Highway Administration.

Flammable Any substance that will easily catch fire or explode.

Flare To spread gradually outward in a bell shape.

FMVSS Acronym for Federal Motor Vehicle Safety Standard.

FMVSS No.121 Acronym for Federal Motor Vehicle Safety Standard No 121.

Foot Valve A foot-operated brake valve that controls air pressure to the service chambers.

Foot-Pound A standard unit of measurement for torque. One foot-pound is the torque obtained by a force of 1 pound applied to a foot long wrench handle.

Franchised Dealership A dealer that has signed a contract with a manufacturer to sell and service a specific line of vehicles.

Fretting A result of vibration that causes the bearing outer race to pick up the machining pattern.

Front Axle Limiting Valve A valve that reduces brake pressure to the front service chambers reducing the chance of front-wheel lock-up on wet or icy pavement.

Full Floating Axles An axle configuration whereby the axle half shafts transmit only driving torque to the wheels and not bending and torsional loads that are characteristic of the semi-floating axle.

Fusible Link A term often used for fuse link.

Fuse Link A short length of smaller gauge wire installed in a conductor, usually close to the power source.

GCW Acronym for gross combination weight.

Gross Combination Weight (GCW) The total weight of a fully quipped vehicle including payload, fuel, and driver.

Gross Vehicle Weight (GVW) The total weight of a fully equipped vehicle and its payload.

Ground The negatively charged side of a circuit. A ground can be a wire, the negative side of the battery, or the vehicle chassis.

Grounded Circuit A shorted circuit that causes current to return to the battery before it has reached its intended destination.

GTW Acronym for gross trailer weight.

GVW Acronym for gross vehicle weight.

Harness and Harness Connectors The connection backbone of the vehicle's electrical system.

Hazardous Materials Any substance that is flammable, explosive, or is known to produce adverse health effects in people or the environment that are exposed to the material during its use.

Heads-Up Display (HUD) A technology used in some vehicles that superimposes data on the driver's normal field of vision. Allows the driver to monitor conditions such as road speed without interrupting his normal view of traffic.

Heat Exchanger A device used to transfer heat, such as a radiator or condenser.

HUD Acronym for heads-up display.

Hydraulic Brakes Brakes actuated by a hydraulic circuit.

Hydraulic Brake System A system using hydraulics to activate the brakes.

I-Beam Axle An axle forged so the cross section of the axle resembles the letter "I."

ICC Check Valve A valve that allows air to flow in one direction only. It is a federal requirement to have a check valve between the wet and dry air tanks.

Inboard Toward the centerline of the vehicle.

In-Line Fuse A fuse that is in series with the circuit.

Installation Templates Drawings supplied by some vehicle manufacturers to allow the technician to correctly install the accessory. Used for auto slack installation.

Insulator A material, such as rubber or glass, that offers high resistance to the flow of electrons.

Integrated Circuit A solid state component containing diodes, transistors, resistors, and capacitors.

Jacobs Engine Brake A hydraulically operated device that converts a power producing diesel engine into a power-absorbing retarder mechanism.

Jumper Wire A wire used to temporarily bypass a circuit or components for electrical testing. A jumper wire consists of a length of wire with an alligator clip at each end.

Kinetic Energy Energy in motion.

Lateral Run Out The wobble or side-to-side movement of a rotating wheel.

Lazer Beam Alignment System A two- or four-wheel alignment system using wheel-mounted instruments to project a lazer beam to measure toe, caster, and camber.

Linkage A system of rods and levers used to transmit motion or force.

Live Axle Axles that drive the wheels.

Load Proportioning Valve (LPV) Valve used to redistribute hydraulic pressure to front and rear brakes based on vehicle loads. This is a load- or height-sensing valve that senses the vehicle load and proportions braking between front and rear brakes in proportion to load variations and weight transfer.

Lockstrap A manual adjustment mechanism that allows for the adjustment of free travel.

Maintenance Manual A publication containing routine maintenance procedures and intervals for vehicle components and systems.

Metering Valve Valve used on vehicles equipped with front disc and rear drum brakes. It improves braking balance during light brake applications by preventing application of the front disc brakes until pressure is built up in the rear brake system.

Moisture Ejector Valve mounted to the bottom of supply and service reservoirs that expels water.

NATEF Acronym for National Automotive Technicians Education Foundation.

National Automotive Technicians Education Foundation (NATEF) A foundation having a program of certifying secondary and post secondary automotive and heavy-duty truck training programs.

National Institute for Automotive Service Excellence (ASE) A nonprofit organization that has an established certification program for school bus automotive, heavy-duty truck, auto body repair, engine machine shop technicians, and parts specialists.

NIASE Acronym for National Institute for Automotive Service Excellence, now abbreviated ASE.

NIOSH Acronym for National Institute for Occupation Safety and Health.

NLGI Acronym for National Lubricating Grease Institute.

NHTSA Acronym for National Highway Traffic Safety Administration.

OEM Acronym for original equipment manufacturer.

Off-Road Unpaved, rough, or ungraded terrain. Any terrain not considered part of the highway system falls into this category.

Ohm A unit of measured electrical resistance.

Ohm's Law Law of electricity stating that in any electrical circuit, current, resistance, and pressure work together in a mathematical relationship.

On-road Paved or smooth-graded surface terrain generally considered to be part of the public highway system.

Open Circuit An electrical circuit whose path has been interrupted or broken either accidentally (a broken wire) or intentionally (a switch turned off).

Operational Control Valve A valve used to control the flow of compressed air through the brake system.

Oscillation Cyclical movement or vibration.

OSHA Acronym for Occupational Safety and Health Administration.

Out-of-Round An eccentric bore.

Output Yoke A shaft connecting link to which a U-joint is attached.

Pad A disc brake friction facing.

Parallel Circuit Electrical circuit that provides two or more paths for the current flow.

Parking Brake A mechanically applied brake used to prevent a parked vehicle's movement.

Parts Requisition Form used to order new parts, on which the technician writes the part(s) that are needed along with the vehicle's VIN.

Payload Weight of the cargo carried by a vehicle, not including the weight of the body.

Pitting Surface irregularities resulting from corrosion.

Plies Layers of rubber-impregnated fabric that make up the body of a tire.

Polarity State or charge differential, either positive or negative, in an electrical circuit.

Pounds per Square Inch (psi) A unit of English measure for pressure.

Power A measure of work done factored with time.

Pressure Force applied to a definite area measured in pounds per square inch (psi) English or kilopascals (kPa) metric.

Pressure Differential The difference in pressure between any two points of a system or a component.

Pressure Relief Valve Valve located on the supply tank, usually preset at 150 psi (1,034 kPa). Limits system pressure if the compressor or governor unloader valve malfunctions.

Proportioning Valve Valve used on vehicles equipped with front disc and rear drum brakes. It is installed in the lines to the rear drum brakes, and in a split system, below the pressure differential valve. By reducing pressure to the rear drum brakes, the valve helps to prevent lock-up during severe brake application and provides better braking balance.

psi Acronym for pounds per square inch.

Quick Release Valve Device used to exhaust air close to the service chambers or spring brakes.

Radial A tire design having cord materials running in a direction from the center point of the tire, usually from bead to bead.

Radial Load A load that is applied at 90° to an axis of rotation.

Rated Capacity The maximum, recommended safe load that can be sustained by a component or an assembly without permanent damage.

Ratio Valve Valve used on the steering axle of a some air brake equipped vehicles to modulate brake application pressure chambers during normal service braking.

RCRA Acronym for Resource Conservation and Recovery Act.

Recall Bulletin Bulletin that pertains to service work or replacement of parts in connection with a recall notice.

Reference Voltage The voltage supplied to a sensor by the computer, which acts as a base line voltage; modified by the sensor to act as an input signal.

Relay An electric switch that allows a small current to control a larger one. Consists of a control circuit and a power circuit.

Relay Valve A pneumatic slave valve that receives a signal from a remote source and uses a local air supply to actuate a brake chamber.

Resistance The opposition to current flow in an electrical circuit.

Resource Conservation and Recovery Act (RCRA) Law that states that after using a hazardous material, it must be properly stored until an approved hazardous waste hauler arrives to take it to a disposal site.

Revolutions per Minute (rpm) The number of complete turns a shaft or wheel makes in one minute.

Rollers Hardware that attaches to the web of brake shoes by means of roller retainers. The rollers, in turn, ride the S-cam profile.

Rotation Term used to describe a gear, shaft, or other device that is turning.

rpm Acronym for revolutions per minute.

Rotor Rotating member of an assembly.

Run Out Deviation or wobble of a shaft or wheel has as it rotates. Run out is measured with a dial indicator.

Screw Pitch Gauge Gauge used to provide an accurate method of checking threads per inch of a nut or bolt.

Sensor Electronic device used to input data to an ECU.

Series Circuit Circuit with only one path for electron flow.

Series/Parallel Circuit Circuit designed so that both series and parallel combinations exist within the same circuit.

Service Bulletin Publication that provides service tips, field repairs, product improvements, and related information of benefit to service personnel.

Service Manual A manual, published by the manufacturer, that contains service and repair information for vehicle systems and components.

Shock Absorber Hydraulic device used to dampen vehicle suspension oscillations.

Short Circuit An undesirable connection between two worn or damaged wires in an electrical circuit.

Solenoid Electromagnet that is used to perform mechanical work, made with coil windings wound around an iron tube.

Solid-State Device Electronic semi-conductor device.

Solid Wires Single-strand conductor.

Solvent A petroleum base substance which dissolves other substances.

Spalling Surface fatigue that occurs when chips, scales, or flakes of metal break off due to fatigue rather than wear.

Specialty Service Shop A shop that specializes in areas such as engine rebuilding, transmission/axle overhauling, brake, air conditioning/heating repairs, and electrical/electronic work.

Specific Gravity The ratio of liquid's mass to an equal volume of distilled water.

Stoplight Switch A pneumatic switch that actuates the brake lights. There are two types: (1) A service stoplight switch located in the service circuit, actuated when the service brakes are applied. (2) An emergency stoplight switch located in the emergency circuit and actuated when a pressure loss occurs.

Stranded Wire Wire that is made up of a number of small solid wires, twisted together, to form a single conductor.

Suspension The linkage system between the axles and vehicle frame.

Suspension Height The distance from a specified point on a vehicle to the road surface when not at curb weight.

Swage To reduce or taper.

Switch Device used to control on/off and direct current flow in a circuit.

System Protection Valve A valve to protect the brake system against an accidental loss of air pressure, buildup of excess pressure, or back-flow and reverse air flow.

Tachometer An instrument that indicates rotating shaft speeds, such as that used to indicate crankshaft rpm.

Tandem A pair.

Time Guide Reference material used for computing compensation payable by the truck manufacturer for repairs or service work to vehicles under warranty.

Torque Twisting force.

Toxicity A statement of how poisonous a substance is.

Tracking The travel of the rear wheels in a parallel path with the front wheels.

Treadle Dual foot brake valve that modulates air from the service reservoirs to service lines and brake chambers.

Treadle Valve A foot-operated brake valve that modulates air pressure to the service chambers.

Tree Diagnosis Chart Chart used to provide a logical sequence for what should be inspected or tested when troubleshooting.

TVW Acronym for (1) Total vehicle weight. (2) Towed vehicle weight.

U-Bolt A fastener used to clamp the top U-bolt plate, spring, axle, and bottom U-bolt plate together. Inverted (nuts down) U-bolts cross springs when in place; conventional (nuts up) U-bolts wrap around the axle.

Underslung Suspension Suspension in which the spring is positioned under the axle.

Universal Joint (U-joint) A component that allows torque to be transmitted to components that are operating at different angles.

Vacuum Absence of matter but often used to describe pressure conditions below atmospheric pressure.

Validity List A list of current valid bulletins.

Vehicle Retarder Auxiliary braking device that supplements the service brakes on commercial vehicles. Includes engine, flywheel, transmission, and driveline braking devices.

VIN Acronym for Vehicle Identification Number.

Viscosity Viscosity describes oil thickness or resistance to flow.

Volt The unit of electromotive force.

Voltage-Generating Sensors Devices which produce their own voltage signal. An ABS wheel speed sensor is an example.

Wedge Brakes Brake system using air pressure and brake chambers to push a wedge and roller assembly into an actuator located between adjusting and anchor pistons.

Wet Tank Commonly used slang for a supply reservoir.

Wheel and Axle Speed Sensors Electromagnetic devices used to monitor vehicle speed information for an antilock controller.

Wheel Balance The equal distribution of weight in a wheel with the tire mounted. It is an important factor which affects tire wear and vehicle control.

Notes

Notes

Notes

Notes

Notes

Notes

Notes

Notes

Notes

Notes

Notes

Notes

Notes

Notes

Notes